More praise for
100 Essential Things You Didn't Know You Didn't Know

D0348647

"Fascinating. . . . Though unpredictably diverse, this treasury piquantly reminds readers of how much we err when we dismiss mathematics as a dryly academic specialty, cut off from the rhythms of real life. . . . Where else does math become a romp, full of entertaining tricks and turns?"   —Bryce Christensen, *Booklist*

"John Barrow comprehensively explodes the myths that mathematics is limited to what we learn at school and that none of it is useful in the real world. This dip-in-anywhere book shows you the hidden mathematics behind diamond cutting, high finance, even standing in a queue. Essential reading for anyone who wants to know why we need mathematics."

—Ian Stewart, author of *Professor Stewart's Cabinet of Mathematical Curiosities*

"The repetition in the title of this book captures its intriguing character: that knowledge about knowledge is itself knowledge, including what we don't know. John Barrow's focus is on the mathematical ideas that offer insight into what makes the world turn around. They crop up in everything from physics to politics and whet one's appetite for digging deeper. . . . [Barrow] directs one's attention to their larger significance, leaving the reader to see the world with a new understanding."

—Steven J. Brams, professor of politics, New York University

BY THE SAME AUTHOR

Theories of Everything

The Left Hand of Creation
(with Joseph Silk)

L'Homme et le Cosmos
(with Frank J. Tipler)

The Anthropic Cosmological Principle
(with Frank J. Tipler)

The World within the World

The Artful Universe

Pi in the Sky

Perchè il mondo è matematico?

Impossibility

The Origin of the Universe

Between Inner Space and Outer Space

The Universe that Discovered Itself

The Book of Nothing

The Constants of Nature:
From Alpha to Omega

The Infinite Book:
A Short Guide to the Boundless,
Timeless and Endless

Cosmic Imagery:
Key Images in the History of Science

100 Essential Things You Didn't Know You Didn't Know

Math Explains Your World

JOHN D. BARROW

W. W. Norton & Company
New York · London

Copyright © 2008 by John D. Barrow
First American Edition 2009

St Louis Gateway Arch (p. 29) © Alamy; Diamonds (p. 214 bottom; p. 215) © SPL.

For information about permission to reproduce selections from this book,
write to Permissions, W. W. Norton & Company, Inc., 500 Fifth Avenue,
New York, NY 10110

For information about special discounts for bulk purchases, please contact
W. W. Norton Special Sales at specialsales@wwnorton.com or 800-233-4830

Manufacturing by Courier Westford
Production manager: Anna Oler

Library of Congress Cataloging-in-Publication Data

Barrow, John D., 1952–
    100 essential things you didn't know you didn't know : math explains your world /
John D. Barrow.—1st American ed.
        p. cm.
    Includes bibliographical references and index.
    ISBN 978-0-393-07007-1 (hardcover)
1. Mathematics—Miscellanea. I. Title. II. Title: One hundred essential things you
didn't know you didn't know.
    QA99.B188 2009
    510—dc22   2008055910

ISBN 978-0-393-33867-6 pbk.

W. W. Norton & Company, Inc.
500 Fifth Avenue, New York, N.Y. 10110
www.wwnorton.com

W. W. Norton & Company Ltd.
Castle House, 75/76 Wells Street, London W1T 3QT

3 4 5 6 7 8 9 0

I continued to do arithmetic with my father, passing proudly through fractions to decimals. I eventually arrived at the point where so many cows ate so much grass, and tanks filled with water in so many hours. I found it quite enthralling.

Agatha Christie

To David and Emma

Contents

Preface

This is a little book of bits and pieces – bits about off-beat applications of mathematics to everyday life, and pieces about a few other things not so very far away from it. There are a hundred to choose from, in no particular order, with no hidden agenda and no invisible thread. Sometimes you will find only words, but sometimes you will find some numbers as well, and very occasionally a few further notes that show you some of the formulae behind the appearances. Maths is interesting and important because it can tell you things about the world that you can't learn in any other way. When it comes to the depths of fundamental physics or the breadth of the astronomical universe we have almost come to expect that. But I hope that here you will see how simple ideas can shed new light on all sorts of things that might otherwise seem boringly familiar or just pass by unnoticed.

Lots of the examples contained in the pages to follow were stimulated by the goals of the Millennium Mathematics Project[1], which I came to Cambridge to direct in 1999. The challenge of showing how mathematics has something to tell about most things in the world around us is one that, when met, can play an important part in motivating and informing people, young and old, to appreciate and understand the place of mathematics at the root of our understanding of the world.

I would like to thank Steve Brams, Marianne Freiberger, Jenny

1 www.mmp.maths.org

Gage, John Haigh, Jörg Hensgen, Helen Joyce, Tom Körner, Imre Leader, Drummond Moir, Robert Osserman, Jenny Piggott, David Spiegelhalter, Will Sulkin, Rachel Thomas, John H. Webb, Marc West, and Robin Wilson for helpful discussions, encouragement, and other practical inputs that contributed to the final collection of essential things you now find before you.

Finally, thanks to Elizabeth, David, Roger and Louise for their unnervingly close interest in this book. Several of these family members now often tell me why pylons are made of triangles and tightrope walkers carry long poles. Soon you will know too.

John D. Barrow
August 2008, Cambridge

1

Two's Company, Three's a Crowd

What goes up must come down.

Anon.

Two people who get on well together can often find their relationship destabilised by the arrival of a third into their orbit. This is even more noticeable when gravity is the force of attraction involved. Newton taught us that two masses can remain in stable orbit around their centre of mass under their mutual gravitational forces – as do the Earth and the Moon. But if a third body of similar mass is introduced into the system, then something quite dramatic generally happens. One body finds itself kicked out of the system by the gravitational forces, while the two that remain are drawn into a more tightly bound stable orbit.

This simple 'slingshot' process is the source of a fantastic counter-intuitive property of Newton's theory of gravity discovered by Jeff Xia in 1992. First, take four particles of equal mass M and arrange them in two pairs orbiting within two planes that are parallel and with opposite directions of spin so there is no overall rotation. Now introduce a fifth much lighter particle m that oscillates back and forth along the perpendicular through the mass centres of the two pairs. The group of five particles will expand to infinite size in a finite time!

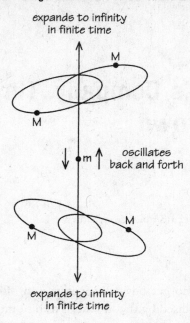

How does this happen? The little oscillating particle runs from one pair to the other, and at the other it creates a little 3-body problem and gets ejected, and the pair recoils outwards to conserve momentum. The lightest particle then travels across to the other pair and the same scenario is repeated. This happens time and time again, without end, and accelerates the two pairs so hard that they become infinitely separated in a finite time, undergoing an infinite number of oscillations in the process.

This example actually solves an old problem posed by philosophers as to whether it is possible to perform an infinite number of actions in a finite time. Clearly, in a Newtonian world where there is no speed limit, it is. Unfortunately (or perhaps fortunately), this behaviour is not possible when Einstein's relativity is taken into account. No information can be transmitted faster than the speed of light and gravitational forces cannot become arbitrarily strong in Einstein's theory of motion and gravitation.

Nor can masses get arbitrarily close to each other and recoil. When two masses of mass M get closer than a distance $4GM/c^2$, where G is Newton's gravitation constant and c is the speed of light, then a 'horizon' surface of no-return forms around them and they form a black hole from which they cannot escape.

The slingshot effect of gravity can be demonstrated in your back garden with a simple experiment. It shows how three bodies can combine to create big recoil as they try to conserve momentum when they pass close to each other (in the case of astronomical bodies) or collide (as it will be in our experiment).

The three bodies will be the Earth, a large ball (like a basket ball or smooth-surfaced football) and a small ball (like a ping-pong or tennis ball). Hold the small ball just above the large ball at about chest height and let them both fall to the ground together. The big ball will hit the ground first and rebound upwards, hitting the small ball while it is still falling. The result is rather dramatic. The small ball bounces up to a height about *nine* times higher than it would have gone if it had just been dropped on the ground from the same height.* You might not want to do this indoors!

---

* The basket ball rebounds from the ground with speed V and hits the ping-pong ball when it is still falling at speed V. So, relative to the basket ball, the ping-pong ball rebounds upwards at speed 2V after its velocity gets reversed by the collision. Since the basket ball is moving at speed V relative to the ground this means that the ping-pong ball is moving upwards at 2V + V = 3V relative to the ground after the collision. Since the height reached is proportional to V^2 this means that it will rise $3^2 = 9$ times higher than in the absence of its collision with the basket ball. In practice, the loss of energy incurred at the bounces will ensure that it rises a little less than this.

2

It's a Small World After All

It's a small world but we all run in big circles.

Sasha Azevedo

How many people do you know? Let's take a round number, like 100, as a good average guess. If your 100 acquaintances each know another 100 different people then you are connected by one step to 10,000 people: you are actually better connected than you thought. After n of these steps you are connected to $10^{2(n+1)}$ people. As of this month the population of the world is estimated to be 6.65 billion, which is $10^{9.8}$, and so when $2(n+1)$ is bigger than 9.8 you have more connections than the whole population of the world. This happens when n is bigger than 3.9, so just 4 steps would do it.

This is a rather remarkable conclusion. It makes a lot of simple assumptions that are not quite true, like the one about all your friend's friends being different from each other. But if you do the counting more carefully to take that into account, it doesn't make much difference. Just six steps is enough to put you in contact with just about anyone on Earth. Try it – you will be surprised how often fewer than 6 steps links you to famous people.

There is a hidden assumption in this that is not well tested by thinking about your links to the Prime Minister, David Beckham or the Pope. You will be surprisingly close to these famous people because they have many links with others. But try linking to an

Amazonian Indian tribe member or a Mongolian herdsman and you will find that the chain of links is much longer. You might not even be able to close it. These individuals live in 'cliques', which have very simple links once you look beyond their immediate close-knit communities.

$$X\text{-----}X\text{-----}X\text{-----}X\text{-----}X$$

If your network of connections is a chain or a circle, then you are only connected by one person on either side and overall connectedness is poor. However, if you are in a ring of these connections with some other random links added, then you can get from one point on the ring to any other very quickly.

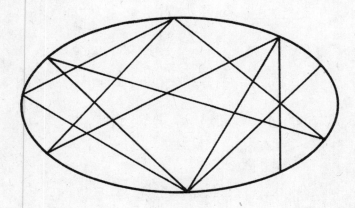

In recent years we have begun to appreciate how dramatic the effects of a few long-range connections can be for the overall connectivity. A group of hubs that produce lots of connections to nearby places can be linked up very effectively by just adding a few long-range connections between them.

These insights are important when it comes to figuring out how much coverage you need in order for mobile phone links between all users to be possible, or how a few infected individuals can spread a disease by their interactions with members of a population. When airlines try to plan how they arrange hubs and routes so as to minimise journey times, or maximise the number of cities they can connect with a single change of planes or minimise costs, they need to understand the unexpected properties of these 'small world' networks.

The study of connectedness shows us that the 'world' exists at many levels: transport links, phone lines, email paths all create networks of interconnections that bind us together in unlikely ways. Everything is rather closer than we thought.

3

Monkey Business

I have a spelling chequer
It came with my pee sea
It plainly marques four my revue
Miss takes I cannot see

I've run this poem threw it
I'm shore yaw pleased to no
It's letter perfect in its weigh
My chequer told me sew . . .

Barri Haynes

The legendary image of an army of monkeys typing letters at random and eventually producing the works of Shakespeare seems to have emerged gradually over a long period of time. In *Gulliver's Travels*, written in 1782, Jonathan Swift tells of a mythical Professor of the Grand Academy of Lagado who aims to generate a catalogue of all scientific knowledge by having his students continuously generate random strings of letters by means of a mechanical printing device. The first mechanical typewriter had been patented in 1714. After several eighteenth- and nineteenth-century French mathematicians used the example of a great book being composed by a random deluge of letters from a printing works as an example of extreme improbability, the monkeys appear first in 1909, when the French mathematician Émile Borel suggested that randomly typing monkeys would eventually produce every book in France's Bibliothèque Nationale. Arthur Eddington took up the analogy in his famous book *The Nature of the Physical World* in 1928, where he anglicised the library: 'If I let my fingers wander idly over the keys of a typewriter it *might* happen that my screed made an intelligible sentence. If an army of monkeys were strumming on typewriters they *might* write all the books in the British Museum.'

Eventually this oft-repeated example picked the 'Complete Works of Shakespeare' as the prime candidate for random recreation. Intriguingly, there was a website that once simulated an ongoing random striking of typewriter keys and then did pattern searches against the 'Complete Works of Shakespeare' to identify matching character strings. This simulation of the monkeys' actions began on 1 July 2003 with 100 monkeys, and the population of monkeys was effectively doubled every few days until recently. In that time they produced more than 10^{35} pages, each requiring 2,000 keystrokes.

A running record was kept of daily and all-time record strings until the Monkey Shakespeare Simulator Project site stopped updating in 2007. The daily records are fairly stable, around the 18- or 19-character-string range, and the all-time record inches steadily upwards. For example, one of the 18-character strings that the monkeys have generated is contained in the snatch:

. . . Theseus. Now faire UWfllaNWSK2d6L;wb . . .

The first 18 characters match part of an extract from *A Midsummer Night's Dream* that reads

. . . us. Now faire Hippolita, our nuptiall houre . . .

For a while the record string was 21-characters long, with

. . . KING. Let fame, that wtIA'"yh!"VYONOvwsFOsbhzkLH . . .

which matches 21 letters from *Love's Labour's Lost*

> KING. Let fame, that all hunt after in their lives,
> Live regist'red upon our brazen tombs,
> And then grace us in the disgrace of death; . . .

In December 2004 the record reached 23 characters with

> Poet. Good day Sir FhlOiX5a]OM,MLGtUGSxX4IfeHQbktQ . . .

which matched part of *Timon of Athens*

> Poet. Good day Sir
> Pain. I am glad y'are well
> Poet. I haue not seene you long, how goes the World?
> Pain. It weares sir, as it growes . . .

By January 2005, after 2,737,850 million billion billion billion monkey-years of random typing, the record stretched to 24 characters, with

> RUMOUR. Open your ears; 9r"5j5&?OWTY Z0d 'B-nEoF.vjSqj[. . .

which matches 24 letters from *Henry IV Part 2*

> RUMOUR. Open your ears; for which of you will stop
> The vent of hearing when loud Rumour speaks?

Which all goes to show: it is just a matter of time!

4

Independence Day

I read that there's about 1 chance in 1000 that someone will board an airplane carrying a bomb. So I started carrying a bomb with me on every flight I take; I figure the odds against two people having bombs are astronomical.

Anon.

Independence Day, 4 July 1977 is a date I remember well. Besides being one of the hottest days in England for many years, it was the day of my doctoral thesis examination in Oxford. Independence, albeit of a slightly different sort, turned out to be of some importance because the first question the examiners asked me wasn't about cosmology, the subject of the thesis, at all. It was about statistics. One of the examiners had found 32 typographical errors in the thesis (these were the days before word-processors and schpel-chequers). The other had found 23. The question was: how many more might there be which neither of them had found? After a bit of checking pieces of paper, it turned out that 16 of the mistakes had been found by both of the examiners. Knowing this information, it is surprising that you can give an answer as long as you assume that the two examiners work independently of each other, so that the chance of one finding a mistake is not affected by whether or not the other examiner finds a mistake.

Let's suppose the two examiners found A and B errors respectively and that they found C of them in common. Now assume that the first examiner has a probability a of detecting a mistake while the other has a probability b of detecting a mistake. If the total number of typographical errors in the thesis was T, then A $= aT$ and $B = bT$. But if the two examiners are proofreading *independently* then we also know the key fact that $C = abT$. So $AB = abT^2 = CT$ and so the total number of mistakes is $T = AB/C$, irrespective of the values of a and b. Since the total number of mistakes that the examiners found (noting that we mustn't double-count the C mistakes that they both found) was $A + B - C$, this means that the total number that they didn't spot is just $T - (A + B - C)$ and this is $(A - C)(B - C)/C$. In other words, it's the product of the number that each found that the other didn't divided by the number they both found. This makes good sense. If both found lots of errors but none in common then they are not very good proofreaders and there are likely to be many more that neither of them found. In my thesis we had $A = 32$, $B = 23$, and $C = 16$, so the number of unfound errors was expected to be $(16 \times 7)/16 = 7$.

This type of argument can be used in many situations. Suppose different oil prospectors search independently for oil pockets: how many might lie unfound? Or if ecologists want to know how many animal or bird species might there be in a region of forest if several observers do a 24-hour census.

A similar type of problem arose in literary analysis. In 1976 two Stanford statisticians used the same approach to estimate the size of William Shakespeare's vocabulary by investigating the number of different words used in his works, taking into account multiple usages. Shakespeare wrote about 900,000 words in total. Of these, he uses 31,534 different words, of which 14,376 appear only once, 4,343 appear only twice and 2,292 appear only three times. They predict that Shakespeare knew at least 35,000 words that are not used in his works: he probably had a total vocabulary of about 66,500 words. Surprisingly, you know about the same number.

5

Rugby and Relativity

Rugby football is a game I can't claim absolutely to understand in all its niceties, if you know what I mean. I can follow the broad, general principles, of course. I mean to say, I know that the main scheme is to work the ball down the field somehow and deposit it over the line at the other end and that, in order to squalch this programme, each side is allowed to put in a certain amount of assault and battery and do things to its fellow man which, if done elsewhere, would result in 14 days without the option, coupled with some strong remarks from the Bench.

P.G. Wodehouse, *Very Good, Jeeves*

Relativity of motion need not be a problem only for Einstein. Who has not had the experience of sitting in a stationary railway carriage at a station, then suddenly getting the sensation of being in motion, only to recognise that a train on the parallel track has just moved off in the other direction and your train is not moving at all?

Here is another example. Five years ago I spent two weeks visiting the University of New South Wales in Sydney during the time that the Rugby World Cup was dominating the news media and public interest. Watching several of these games on television I noticed an interesting problem of relativity that was unnoticed by the celebrities in the studio. What is a forward pass relative to? The written rules are clear: a forward pass occurs when the ball is thrown towards the opposing goal line. But when the players are moving the situation becomes more subtle for an observer to judge due to relativity of motion.

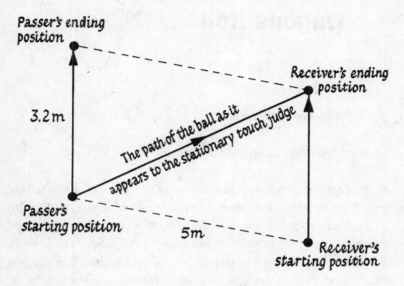

Imagine that two attacking players are running (up the page) in parallel straight lines 5 metres apart at a speed of 8 metres per sec towards their opponents' line. One player, the 'receiver', is a metre behind the other, the 'passer', who has the ball. The passer throws the ball at 10 metres per sec towards the receiver. The speed of the ball relative to the ground is actually $\sqrt{(10^2 + 8^2)} = 12.8$ metres per sec and it takes a time of 0.4 sec to travel the 5 metres between the players. During this interval the receiver has run a further distance of $8 \times 0.4 = 3.2$ metres. When the pass was thrown he was 1 metre behind the passer but when he catches the ball he is 2.2 metres in front of him from the point of view of a touch judge standing level with the original pass. He believes there has been a forward pass and waves his flag. But the referee is running alongside the play, doesn't see the ball go forwards, and so waves play on!

6

Wagons Roll

My heart is like a wheel.

Paul McCartney, 'Let Me Roll It'

One weekend I noticed that the newspapers were discussing proposals to introduce more restrictive speed limits of 20 mph in built-up areas of the UK and to enforce them with speed cameras wherever possible. Matters of road safety aside, there are some interesting matters of rotational motion that suggest that speed cameras might end up catching large numbers of perplexed cyclists apparently exceeding the speed limit by significant factors. How so?

Suppose that a cycle is moving at speed V towards a speed detector. This means that a wheel hub or the body of the cyclist is moving with speed V with respect to the ground. But look more carefully at what is happening at different points of the spinning wheel. If the wheel doesn't skid, then the speed of the point of the wheel in contact with the ground must be zero. If the wheel has radius R and is rotating with constant angular velocity Ω revolutions per second, then the speed of the contact point can also be written as $V - R\,\Omega$. This must be zero and therefore V equals $R\,\Omega$. The forward speed of the centre of the wheel is V, but the forward speed of the top of the wheel is the sum of V and the rotational speed. This equals $V + R\,\Omega$ and is therefore equal to 2V. If a camera determines the speed of an approaching or receding bicycle by measuring the speed of the top of the wheel, then it will register a speed twice as large as the cyclist is moving. An interesting one for m'learned friends perhaps, but I recommend you have a good pair of mudguards.

7

A Sense of Proportion

You can only find truth with logic if you have already found truth without it.

G.K. Chesterton

As you get bigger, you get stronger. We see all sorts of examples of the growth of strength with size in the world around us. The superior strength of heavier boxers, wrestlers and weightlifters is acknowledged by the need to grade competitions by the weight of the participants. But how fast does strength grow with increasing weight or size? Can it keep pace? After all, a small kitten can hold its spiky little tail bolt upright, yet its much bigger mother cannot: her tail bends over under its own weight.

Simple examples can be very illuminating. Take a short bread-stick and snap it in half. Now do the same with a much longer one. If you grasped it at the same distance from the snapping point each time you will find that it is no harder to break the long stick than to break the short one. A little reflection shows why this should be so. The stick breaks along a slice through the bread-stick. All the action happens there: a thin sheet of molecular bonds in the breadstick is broken and it snaps. The rest of the breadstick is irrelevant. If it were a hundred metres long it wouldn't make it any harder to break that thin slice of bonds at one point along its length. The strength of the breadstick is given by the number of molecular bonds that have to be broken across its cross-sectional area. The bigger that area, the more bonds that need to be broken and the stronger the breadstick. So strength is proportional to cross-sectional area, which is usually proportional to some measure of its diameter squared.

Everyday things like breadsticks and weightlifters have a constant density that is just determined by the average density of the atoms that compose them. But density is proportional to mass divided by volume, which is mass divided by the cube of size. Sitting here on the Earth's surface, mass is proportional to weight, and so we expect the simple proportionality 'law' that for fairly spherical objects

$$(\text{Strength})^3 \propto (\text{weight})^2$$

This simple rule of thumb allows us to understand all sorts of things. The ratio of strength to weight is seen to fall as Strength/Weight $\propto$ (weight)$^{-1/3}$ $\propto$ 1/(size) . So as you grow bigger, your strength does not keep pace with your increasing weight. If all your dimensions expanded uniformly in size, you would eventually be too heavy for your bones to support and you would break. This is why there is a maximum size for land-going structures made of atoms and molecules, whether they are dinosaurs, trees or buildings. Scale them up in shape and size and eventually they will grow so big that their weight is sufficient to sever the molecular bonds at their base and they will collapse under their own weight.

We started by mentioning some sports where the advantage of size and weight is so dramatic that competitors are divided into different classes according to their bodyweight. Our 'law' predicts that we should expect to find a straight-line correlation when we plot the cube of the weight lifted against the square of the body-weight of weightlifters. Here is what happens when you plot that graph for the current men's world-records in the clean and jerk across the weight categories:

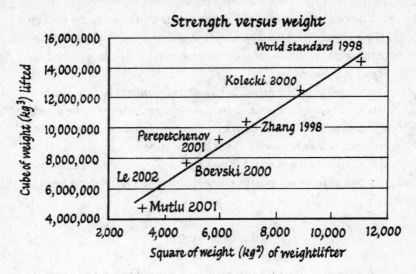

Strength versus weight

It's an almost perfect fit! Sometimes mathematics can make life simple. The weightlifter who lies farthest above the line giving the 'law' is the strongest lifter 'pound for pound', whereas the heaviest lifter, who lifts the largest weight, is actually relatively the weakest when his size is taken into account.

8

Why Does the Other Queue Always Move Faster?

The other man's grass is always greener.
The sun shines brighter on the other side.

Sung by Petula Clark

You will have noticed that when you join a queue at the airport or the post office, the other queues always seem to move faster. When the traffic is heavy on the motorway, the other lanes always seem to move faster than the one you chose. Even if you change to one of the others, it still goes slower. This situation is often known as 'Sod's Law' and appears to be a manifestation of a deeply antagonistic principle at the heart of reality. Or, perhaps it is merely another manifestation of human paranoia or a selective recording of evidence. We are impressed by coincidences without pausing to recall all the far more numerous non-coincidences we never bothered to keep a note of. In fact, the reason you so often seem to be in the slow queue may not be an illusion. It is a consequence of the fact that on the average you *are* usually in the slow queue!

The reason is simple. On the average, the slow lines and lanes are the ones that have more people and vehicles in them. So, you are more likely to be in those, rather than in the faster moving ones where fewer people are.

The proviso 'on the average' is important here. Any particular queue will possess odd features – people who forgot their wallet, have a car that won't go faster than 30 mph and so on. You won't invariably be in the slowest line, but *on the average*, when you consider all the lines that you join, you will be more likely to be in the more crowded lines where most people are.

This type of self-selection is a type of bias that can have far-reaching consequences in science and for the analysis of data, especially if it is unnoticed. Suppose you want to determine if people who attend church regularly are healthier than those who do not. There is a pitfall that you have to avoid. The most unhealthy people will not be able to get to church and so just counting heads in the congregation and noting their state of health will give a spurious result. Similarly, when we come to look at the Universe we might have in mind a 'principle', inspired by Copernicus, that we must not think that our position in the Universe is special. However, while we should not expect our position to be special in *every* way, it would be a grave mistake to believe that it cannot be special in *any* way. Life may be possible only in places where special conditions exist: it is most likely to be found where there are stars and planets. These structures form in special places where the abundance of dusty raw material is higher than average. So, when we do science or are confronted with data the most important question to ask about the results is always whether some bias is present that leads us preferentially to draw one conclusion rather than another from the evidence.

9

Pylon of the Month

Like Moses parting the waves, National Grid Company PLC's 4YG8 leads his fellow pylons through this Oxfordshire housing estate towards the 'promised land' of Didcot Power Station.

The December 1999 *Pylon of the Month*

There are some fascinating websites about, but none was more beguiling than the iconic *Pylon of the Month*,* once devoted to providing monthly pin-ups of the world's most exciting and seductive electricity pylons. The ones shown on the website below are from Scotland. Alas, *Pylon of the Month* now seems to have become a cobweb site, but there is still something to learn from it, since for the mathematician every pylon tells a story. It is about something so prominent and ubiquitous that, like gravity, it goes almost unnoticed.

Next time you go on a train journey, look carefully at the pylons as they pass swiftly by the windows. Each is made of a network of metal struts that make use of a single recurring polygonal shape. That shape is the triangle. There are big triangles and smaller ones nested within them. Even apparent squares and rectangles are merely separate pairs of triangles. The reason forms a small part of an interesting mathematical story that began in the early nineteenth century with the work of the French mathematician Augustin-Louis Cauchy.

* http://www.drookitagain.co.uk/coppermine/thumbnails.php?album=34

Of all the polygonal shapes that we could make by bolting together straight struts of metal, the triangle is special. It is the only one that is *rigid*. If they were hinged at their corners, all the others can be flexed gradually into a different shape without bending the metal. A square or a rectangular frame provides a simple example: we see that it can be deformed into a parallelogram without any buckling. This is an important consideration if you aim to maintain structural stability in the face of winds and temperature changes. It is why pylons seem to be great totems to the god of all triangles.

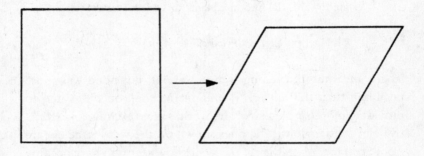

If we move on to three-dimensional shapes then the situation is quite different: Cauchy showed that *every* convex polyhedron (i.e. in which the faces all point outwards) with rigid faces, and hinged along its edges, is rigid. And, in fact, the same is true for convex polyhedra in spaces with four or more dimensions as well.

What about the non-convex polyhedra, where some of the faces can point inwards? They look much more squashable. Here, the question remained open until 1978 when Robert Connelly found an example with non-convex faces that is not rigid and then showed that in all such cases the possible flexible shifts keep the total volume of the polyhedron the same. However, the non-convex polyhedral examples that exist, or that may be found in the future, seem to be of no immediate practical interest to structural engineers because they are special in the sense that they require a perfectly accurate construction, like balancing a needle on its point. Any deviation from it at all just gives a rigid example, and so mathematicians say that 'almost every' polyhedron is rigid. This all seems to make structural stability easy to achieve – but pylons do buckle and fall down. I'm sure you can see why.

10

A Sense of Balance

Despite my privileged upbringing, I'm quite well-balanced. I
have a chip on both shoulders.

Russell Crowe in *A Beautiful Mind*

Whatever you do in life, there will be times when you feel you
are walking a tightrope between success and failure, trying to
balance one thing against another or to avoid one activity gobbling
up every free moment of your time. But what about the people
who really are walking a tightrope. The other day I was watching
some old newsreel film of a now familiar sight: a crazy tightrope
walker making a death-defying walk high above a ravine and a
rushing river. One slip and he would have become just another
victim of Newton's law of gravity.

We have all tried to balance on steps or planks of wood at times,
and we know from experience that some things help to keep you
balanced and upright: don't lean away from the centre, stand up
straight, keep your centre of gravity low. All the things they teach
you in circus school. But those tightrope walkers always seem to
carry very long poles in their hands. Sometimes the poles flop
down at the ends because of their weight, sometimes they even
have heavy buckets attached. Why do you think the funambulists
do that?

The key idea you need to understand why the tightrope walker
carries a long pole to aid balance is inertia. The larger your inertia,

the slower you move when a force is applied. It has nothing to do with centre of gravity. The farther away from the centre that mass is distributed, the higher a body's inertia is, and the harder it is to move it. Take two spheres of different materials that have the same diameter and mass, one solid and one hollow, and it will be the hollow one with all its mass far away at its surface that will be slower to move or to roll down a slope. Similarly, carrying the long pole increases the tightrope walker's inertia by placing mass far away from the body's centre line – inertia has units of mass times distance squared. As a result, any small wobbles about the equilibrium position happen more slowly. They have a longer time period of oscillation, and the walker has more time to respond to the wobbles and restore his balance. Compare how much easier it is to balance a one-metre stick on your finger compared with a 10-centimetre one.

11

Bridging That Gap

Like a bridge over troubled water.

Paul Simon and Art Garfunkel

One of the greatest human engineering achievements has been the construction of bridges to span rivers and gorges that would otherwise be impassable. These vast construction projects often have an aesthetic quality about them that places them in the first rank of modern wonders of the world. The elegant Golden Gate Bridge, Brunel's remarkable Clifton Suspension Bridge and the Ponte Hercilio Luz in Brazil have spectacular shapes that look smooth and similar. What are they?

There are two interesting shapes that appear when weights and chains are suspended and they are often confused or simply assumed to be the same. The oldest of these problems was that of describing the shape that is taken up by a hanging chain or rope whose ends are fixed at two points on the same horizontal level. You can see the shape easily for yourself. The first person to claim they knew what this shape would be was Galileo, who in 1638 maintained that a chain hanging like this under gravity would take up the shape of a parabola (this has the graph $y = Ax^2$ where A is any positive number). But in 1669 Joachim Jungius, a German mathematician who had special interests in the applications of mathematics to physical problems, showed him to be wrong. The determination of the actual equation for the hanging chain was finally calculated by Gottfried Leibniz, Christiaan Huygens, David Gregory and Johann Bernoulli in 1691 after the problem had been publicly announced as a challenge by Johann Bernoulli a year earlier. The curve was first called the *catenaria* by Huygens in a letter to Leibniz, and it was derived from the Latin word *catena* for 'chain', but the introduction of the anglicised equivalent 'catenary' seems to have been due to the US President Thomas Jefferson in a letter to Thomas Paine, dated 15 September 1788, about the design of a bridge. Sometimes the shape was also known as the *chainette* or *funicular* curve.

The shape of a catenary reflects the fact that its tension supports the weight of the chain itself and the total weight born at any point is therefore proportional to the total length of chain between that point and the lowest point of the chain. The equation for the hanging chain has the form $y = B\cosh(x/B)$ where B is the constant tension of the chain divided by its weight per unit length.[1] If you hold two ends of a piece of hanging chain and move them towards each other, or apart, then the shape of the string will continue to be described by this formula but with a different value of B for each position. This curve can also be derived by asking for the shape that makes the centre of gravity of the suspended chain as low as possible.

Another spectacular example of a man-made catenary can be seen in St Louis, Missouri, where the Gateway Arch is an upside-down catenary (see p 29). This is the optimal shape for a self-supporting arch, which minimises the shear stresses because the stress is always directed along the line of the arch towards the ground. Its exact mathematical formula is written inside the arch. For these reasons, catenary arches are often used by architects to optimise the strength and stability of structures; a notable example is in the soaring high arches of Antoni Gaudí's unfinished Sagrada Familia Church in Barcelona.

Another beautiful example is provided by the Rotunda building designed by John Nash in 1819 to be the Museum of Artillery, located on the edge of Woolwich Common in London. Its distinctive tent-like roof, influenced by the shape of soldiers' bell tents, has the shape of one half of a catenary curve.

John Nash's Rotunda building

There is, however, a big difference between a hanging chain and a suspension bridge like the Clifton or the Golden Gate. Suspension bridges don't only have to support the weight of their own cables or chains. The vast bulk of the weight to be supported by the bridge cable is the deck of the bridge. If the deck is horizontal with a constant density and cross-sectional area all the way along it, then the equation for the shape of the supporting cable is now a parabola $y = x^2/2B$, where B is (as for the hanging chain equation) a constant equal to the tension divided by the weight per unit length of the bridge deck.

One of the most remarkable is the Clifton Suspension Bridge

in Bristol, designed by Isambard Kingdom Brunel in 1829 but completed only in 1865, three years after his death. Its beautiful parabolic form remains a fitting monument to the greatest engineer since Archimedes.

The St Louis Gateway Arch

© Alamy

12

On the Cards

Why don't children collect things anymore? Whatever
happened to those meticulously kept stamp albums . . . ?

Woman's Hour, BBC Radio 4

Last weekend, hidden between books in the back of my book-
case, I came across two sets of cards that I had collected as a young
child. Each set contained fifty high-quality colour pictures of classic
motor cars with a rather detailed description of their design and
mechanical specification on the reverse. Collecting sets of cards
was once all the rage. There were collections of wartime aircraft,
animals, flowers, ships and sportsmen – since these collections all
seemed to be aimed at boys – to be amassed from buying lots of
packets of bubble gum, breakfast cereals or packets of tea. Of the
sports cards, just as with today's Panini 'stickers', the favoured
game was football (in the US it was baseball), and I always had
my suspicions about the assumption that all the players' cards were
produced in equal numbers. Somehow everyone seemed to be
trying to get the final 'Bobby Charlton' card that was needed to
complete the set. All the other cards could be acquired by swopping duplicates with your friends, but everyone lacked this essential one.

It was a relief to discover that even my own children engaged
in similar acquisitive practices. The things collected might change
but the basic idea was the same. So what has mathematics got to

do with it? The interesting question is to ask how many cards we should expect to have to buy in order to complete the set, if we assume that each of them is produced in equal numbers and so has an equal chance of being found in the next packet that you open. The motor car sets I came across each contained 50 cards. The first card I get will always be one I haven't got but what about the second card? There is a 49/50 chance that I haven't already got it. Next time it will be a 48/50 chance and so on.

After you have acquired 40 different cards there will be a 10/50 chance that the next one will be one you haven't already got. So on the average you will have to buy another 50/10, or 5 more cards to have a better than evens chance of getting another new one that you need for the set. Therefore, the total number of cards you will need to buy on average to get the whole set of 50 will be the sum of 50 terms:

$$50/50 + 50/49 + 50/48 + \ldots + 50/3 + 50/2 + 50/1$$

where the first term is the certain case of the first card you get and each successive term tells you how many extra cards you need to buy to get the 2nd, 3rd and so on missing members of the set of 50 cards.

As there can be collections with all sorts of different numbers of cards in them, let's consider acquiring a set with any number of cards in it, that we will call N. Then the same logic tells us that on the average we will have to buy a total of

$$(N/N) + (N/N\text{-}1) + (N/N\text{-}2) + \ldots + N/2 + N/1 \text{ cards}$$

Taking out the common factor N in the numerators of each term, this is just

$$N(1 + 1/2 + 1/3 + \ldots + 1/N).$$

The sum of terms in the brackets is the famous 'harmonic' series. When N becomes large it is well approximated by $0.58 + \ln(N)$ where $\ln(N)$ is the natural logarithm of N. So as N gets realistically large we see that the number of cards we need to buy on the average to complete our set is about

$$\text{Cards needed} \approx N \times [0.58 + \ln(N)]$$

For my sets of 50 motor car cards the answer is 224.5, and I should have expected to have to have bought on average about 225 cards to make up my set of 50. Incidentally, our calculation shows how much harder it gets to complete the second half of the collection than the first half. The number of cards that you need to buy in order to collect N/2 cards for half a set is

$$(N/N) + (N/N\text{-}1) + (N/N\text{-}2) + \ldots + N/(\tfrac{1}{2}N+1)$$

which is the difference between N times the harmonic series summed to N and summed to N/2 terms, so

$$\text{Cards needed for half a set} \approx N \times [\ln(N) + 0.58 - \ln(N/2) - 0.58] =$$
$$N\ln(2) = 0.7N$$

Or just 35 to get the first half of my set of 50.

I wonder if the original manufacturers performed such calculations. They should have, because they enable you to work out the maximum possible profit you could expect to gain in the long run from marketing a particular size set of cards. It is likely to be a maximum *possible* profit only because collectors will trade cards and be able to acquire new cards by swopping rather than buying new ones.

What impact can friends make by swopping duplicates with you?

Suppose that you have F friends and you all pool cards in order to build up F+1 sets so that you have one each. How many cards would you need to do this? On the average, when the number of cards N is large, and you share cards, the answer approaches

$$N \times [\ln(N) + F \ln(\ln N) + 0.58]$$

On the other hand, if you had each collected a set without swopping, you would have needed about $(F+1)N[\ln(N) + 0.58]$ cards to complete $F+1$ separate sets. For $N = 50$, the number of card purchases saved would be 156F. Even with $F = 1$ this is a considerable economy.

If you know a little statistics you might like to show that the deviation that can be expected on the $N \times [0.58 + \ln(N)]$ result is close to 1.3N. This is quite significant in practice because it means that you have a 66% chance of needing to collect 1.3 N more or less than the average. For our 50-card set this uncertainty in the expected number of purchases is 65. There was a story a few years ago that a consortium was targeting suitable national lotteries by calculating the average number of tickets that needed to be bought in order to have a good chance of collecting all the possible numbers – and so including the winning one. The members neglected to include the likely variance away from the average result but were very lucky to find that they did have a winning ticket among the millions they had bought.

If the probability of each card appearing is not the same, then the problem becomes harder but is still soluble. In that case it is more like the problem of coin collecting where you try to collect a coin with each available year date on. You don't know whether equal numbers were minted in each year (almost certainly they weren't) or how many may have been withdrawn later, so you can't rely on there being an equal chance of collecting an 1840 penny or an 1890 one. But if you do find a 1933 English penny (of which only 7 were made and 6 are accounted for) then be sure to let me know.

13

Tally Ho

It is a profoundly erroneous truism . . . that we should culti-
vate the habit of thinking of what we are doing. The precise
opposite is the case. Civilization advances by extending the
number of important operations which we can perform
without thinking about them.

Alfred North Whitehead

The hopeless prisoner is incarcerated in a dark, dank, forgotten
cell. The days, months and years are passing slowly. There are
many more to come. The scene is familiar from the movies. But
there is usually an interesting mathematical sub-text. The prisoner
has been keeping track of the days by systems of marks on the
cell wall. The oldest human artefacts with records of counting in
Europe and Africa go back more than 30,000 years and show
similar elaborate groups of tally marks following the days of the
month and the accompanying phases of the Moon.

The standard European pattern of tallying goes back to finger
counting and results in sets of 4 vertical(ish) lines | | | | being
completed by a cross slash to signal a score of five. On to the next
set of five ┼┼┼┤ and so on. The vertical bars plus slash denoting
each counted item show ways of counting that predate our formal
counting systems and led to the adoption of the Roman numerals
I, II and III or the simple Chinese rod numeral systems. They are
closely linked to simple methods of finger counting·and make use

of groups of 5 and 10 as bases for the accumulated sets of marks. Ancient tallying systems recorded marks on bone or were notches carved in wood. Single notches recorded the first few, but a half-cross notch V was then used to mark the 5, with a complete cross X for the ten, hence the V and X Roman numerals; 4 was made either by addition as IIII, or by subtraction as IV. Tallying remained a serious official business in England until as late as 1826, with the Treasury using great wooden tally sticks to keep records of large sums entering and leaving the Exchequer. This use is also the source of the multiple meanings of the word 'score', which means to make a mark and to keep count, as well as the quantity 20. The word tally comes from the word for cut, as still in 'tailor'. When a debt was owed to the Treasury, the tally stick with its scored marks was cleft down the middle and the debtor given one half, the Treasury the other. When the debt was settled the two pieces were joined to check that they 'tallied'.

Counting these tally marks is rather laborious, especially if large totals arise. Individual marks have to be mentally counted and then the number of sets has to be totalled as well. In South America we find the occasional use of a memorable system that uses a gradual build-up of numbers by lines around a square, completed by the two diagonals ⊠.

We are familiar with a variant of this square frame counting when we keep score in a cricket match by making six marks in three rows of two – a 'dot' if no run is scored, or the number scored, or a 'w' if a wicket falls. Other symbols denote wides, byes, no-balls and leg byes. If no runs are scored off all six balls, the six dots are joined up to create the sides of an M, denoting a 'maiden' over; if no runs are scored and a wicket is taken, they are joined to form a W, to indicate a 'wicket maiden' over. In this way a glance at the score book reveals the pattern and number of runless overs.

It is tempting to combine the insights from South American games with those of the cricket scorer to create an ideal aide-memoire for anyone wanting to tally in tens and not have to count up to ten vertical marks to 'see' where the running total has reached. First count from 1 to 4 by placing dots at the four corners of a square, keep on counting from 5 to 8 by adding the four sides, and count 9 and 10 by adding the two diagonals. Each set of 10 is represented by the completed square of four dots and 6 lines. For the next 10 move on to a new square. Completion of a ten is obvious at a glance.

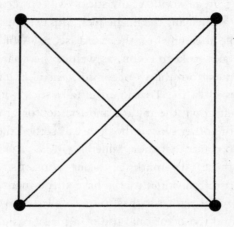

14

Relationships

Relationship: The civilised conversationalist uses this word in public only to describe a seafaring vessel carrying members of his family.

Cleveland Amory

Most magazines have endless articles and correspondence about relationships. Why? Answer: Relationships are complicated, sometimes interesting and can often appear unpredictable. This is just the type of situation that mathematics can help you with.

The simplest relationships between things have a property that we call 'transitivity' and it makes life simple. Being 'taller than' is one of these transitive relationships. So if Ali is taller than Bob and Bob is taller than Carla, then Ali is necessarily taller than Carla. This relationship is a property of heights. But not all relations are like this. Ali might like Bob and Bob might like Carla, but that does not mean that Ali likes Carla. These 'intransitive' relations can create very unusual situations when it comes to deciding what you should do when everyone does not agree.

Suppose that Ali, Bob and Carla decide to invest together in a second-hand car and go out to look at three different possibilities: an Audi, a BMW and a Reliant Robin. They don't all agree on what to buy, so they decide that the outcome must be decided democratically. They must vote on it. So, each of them writes down their order of preference for the three makes:

	First choice	Second choice	Third choice
Ali	Audi	BMW	Reliant Robin
Bob	BMW	Reliant Robin	Audi
Cara	Reliant Robin	Audi	BMW

The voting looks promising at first: Audi beats BMW by 2 preferences to 1, and BMW beats the Reliant Robin by 2 preferences to 1. But, strangely, the Reliant Robin beats the Audi by 2 preferences to 1. Preferring, like 'admiring', is an intransitive relationship that can create awkward paradoxes if not used with care. Small elections to decide who you prefer among candidates for a job, who captains a sports team or even what car to buy are fraught with paradox. Let the voter beware.

Faced with this trilemma, Ali, Bob and Carla decided to give up on the car buying and put their resources into renting a house together. Alas, more decisions were soon necessary. Should they decorate the living room? Should they tidy the garden? Should they buy a new TV? There was no consensus so they decided to vote 'yes' or 'no' on each of the three issues in turn. Here's what they said:

	Decorate house?	Tidy garden?	Buy TV?
Ali	Yes	Yes	No
Bob	No	Yes	Yes
Carla	Yes	No	Yes
Majority decision	YES	YES	YES

All seemed clear. There was a majority of two-to-one to do all three things. All three things should be done. But then money seemed to be running short, and the trio realised that they needed two more people to share the house if they were to pay the rent. After just a few phone calls they had found new house-mates Dell and Tracy, who rapidly moved in with all their belongings. Of course, they thought that it was only fair that they be included in the household vote on the decoration, gardening and TV purchase question. They both voted 'No' to each of the three proposals while Ali, Bob and Carla stuck to their earlier decisions. A very strange situation has now been created in the household.

Here is the table of decisions after Dell and Tracy have added their 'No's:

	Decorate house ?	Tidy garden ?	Buy TV?
Ali	Yes	Yes	No
Bob	No	Yes	Yes
Carla	Yes	No	Yes
Old majority			
decision	*YES*	*YES*	*YES*
Dell	No	No	No
Tracy	No	No	No
Overall decision	*NO*	*NO*	*NO*

We see that their negative votes have tipped the scales on each question. Now there is a majority of 3 to 2 not to decorate the house, not to tidy the garden, and not to buy a TV. But more striking is the fact that a majority (each of Ali, Bob, and Carla) are on the losing side of the vote on two of the three issues. These are the issues on which they don't vote 'No'. So Ali is on the losing side over the house and garden, Bob is on the losing side over the garden and TV, and Carla is on the losing side over the house and the TV. Hence a majority of the people (three out of five) were on the losing side on a majority of the issues (two out of three)!

15

Racing Certainties

'There must be constant vigilance to ensure that any legalised
gambling activity is not penetrated by criminal interests, who
in this connection comprise sophisticated, intelligent, highly
organised, well briefed and clever operators with enormous
money resources which enable them to hire the best brains in
the legal, accountancy, managerial, catering and show business
world.' I am not quite sure about the last two; nevertheless,
that is as true now as it was then.

Viscount Falkland quoting from the *Report of Rothschild
Commission on Gambling (1979)*

A while ago I saw a TV crime drama that involved a plan to
defraud bookmakers by nobbling the favourite for a race. The
drama centred around other goings on, like murder, and the basis
for the betting fraud was never explained. What might have been
going on?

Suppose that you have a race where there are published odds
on the competitors of a_1 to 1, a_2 to 1, a_3 to 1 and so on, for any
number, N, of runners in the race. If the odds are 5 to 4, then we
express that as an a_i of 5/4 to 1. If we lay bets on all of the N
runners in proportion to the odds so that we bet a fraction $1/(a_i
+1)$ of the total stake money on the runner with odds of a_i to 1,
then we will always show a profit as long as the sum of the odds,
which we call Q, satisfies the inequality

$$Q = 1/(a_1 + 1) + 1/(a_2 + 1) + 1/(a_3 + 1) + \ldots + 1/(a_N + 1) < 1$$

And if Q is indeed less than 1, then our winnings will be at least equal to

$$\text{Winnings} = (1/Q - 1) \times \text{our total stake}$$

Let's look at some examples. Suppose there are four runners and the odds for each are 6 to 1, 7 to 2, 2 to 1 and 8 to 1. Then we have $a_1 = 6$, $a_2 = 7/2$, $a_3 = 2$ and $a_4 = 8$ and

$$Q = 1/7 + 2/9 + 1/3 + 1/9 = 51/63 < 1$$

and so by betting our stake money with $1/7$ on runner 1, $2/9$ on runner 2, $1/3$ on runner 3 and $1/9$ on runner 4 we will win at least $51/63$ of the money we staked (and of course we get our stake money back as well).

However, suppose that in the next race the odds on the four runners are 3 to 1, 7 to 1, 3 to 2 and 1 to 1 (i.e. 'evens'). Now we see that we have

$$Q = 1/4 + 1/8 + 2/5 + 1/2 = 51/40 > 1$$

and there is no way that we can guarantee a positive return. Generally, we can see that if there is a large field of runners (so the number N is large) there is a better chance of Q being greater than 1. But large N doesn't necessarily guarantee that we have $Q > 1$. Pick each of the odds by the formula $a_i = i(i+2)$ to 1 and you can get $Q = 3/4$ and a healthy 30 per cent return, even when N is infinite!

But let's return to the TV programme. How is the situation changed if we know ahead of the race that the favourite in our $Q > 1$ example will not be a contender because he has been doped?

If we use this inside doping information we will discount the favourite (with odds of 1 to 1) and place none of our stake money on him. So, we are really betting on a three-horse race where Q is equal to

$$Q = 1/4 + 1/8 + 2/5 = 31/40 < 1$$

and by betting 1/4 of our stake money on runner 1, 1/8 on runner 2 and 2/5 on runner 3 we are guaranteed a minimum return of $(40/31) - 1 = 9/31$ of our total stake in addition to our original stake money! So we are quids in.*

* It has been suggested to me that some criminal money-laundering is performed by spreading on-course bets over all runners, even when Q>1. There is a loss but on average you can predict what it will be and it is just a 'tax' on the money-laundering exchange.

16

High Jumping

Work is of two kinds: first, altering the position of matter at or near the Earth's surface relatively to other such matter; second, telling other people to do so. The first kind is unpleasant and ill paid; the second is pleasant and highly paid.

Bertrand Russell

If you are training to be good at any sport, then you are in the business of optimisation – doing all you can (legally) to enhance anything that will make you do better and minimise any faults that hinder your performance. This is one of the areas of sports science that relies on the insights that are possible by applying a little bit of mathematics. There are two athletics events where you try to launch the body over the greatest possible height above the ground: high jumping and pole vaulting. This type of event is not as simple as it sounds. Athletes must first use their strength and energy to launch their body weight into the air in a gravity-defying manner. If we think of a high jumper as a projectile of mass M launched vertically upwards at speed U, then the height H that can be reached is given by the formula $U^2 = 2gH$, where g is the acceleration due to gravity. The energy of motion of the jumper at take-off is $1/2\,MU^2$ and this will be transformed into the potential energy MgH gained by the jumper at the maximum height H. Equating the two gives $U^2 = 2gH$.

The tricky point is the quantity H – what exactly is it? It is not the height that is cleared by the jumper. Rather, it is the height that the jumper's centre of gravity is raised, and that is a rather subtle thing because it makes it possible for a high jumper's body to pass over the bar even though its centre of gravity passes under the bar.

When an object has a curved shape, like an L, it is possible for its centre of gravity to lie outside the body.* It is this possibility that allows a high jumper to control where his centre of gravity lies and what trajectory it follows when he jumps. The high jumper's aim is to get his body to pass cleanly over the bar while making his centre of gravity pass as far underneath the bar as possible. In this way he will make optimal use of his explosive take-off energy to increase the height cleared.

The simple high-jumping style that you first learn at school, called the 'scissors' technique, is far from optimal. In order to clear the bar your centre of gravity, as well as your whole body, must pass over the bar. In fact your centre of gravity probably goes about 30 centimetres above the height of the bar. This is a very inefficient way to clear a high-jump bar.

The high-jumping techniques used by top athletes are much more elaborate. The old 'straddle' technique involved the jumper rolling around the bar with their chest always facing the bar. This was the favoured technique of world-class jumpers up until 1968 when the American Dick Fosbury amazed everyone by introducing a completely new technique – the 'Fosbury Flop' – which involved a backwards flop over the bar. It won him the Gold Medal at the 1968 Olympics in Mexico City. This method was only safe when

---

* One way to locate the centre of gravity of an object is to hang it up from one point and drop a weighted string from any point on the object, marking where the string drops. Then repeat this by hanging the object up from another point. Draw a second line where the hanging string now falls. The centre of gravity is where the lines of the two strings cross. If the object is a square then the centre of gravity will lie at the geometrical centre but if it is L-shaped or U-shaped the centre of gravity will not generally lie inside the boundary of the body.

inflatable landing areas became available. Fosbury's technique was much easier for high jumpers to learn than the straddle and it is now used by every good high jumper. It enables a high jumper to send his centre of gravity well below the bar even though his body curls over and around it. The more flexible you are the more you can curve your body around the bar and the lower will be your centre of gravity. The 2004 Olympic men's high-jump champion Stefan Holm, from Sweden, is rather small (1.81 m) by the standards of high jumpers, but he is able to curl his body to a remarkable extent. His body is very U-shaped at his highest point. He is able to sail over a bar set at 2 m 37 cm, but his centre of gravity goes well below the bar.

When a high jumper runs in to launch himself upwards, he will be able to transfer only a small fraction of his best possible horizontal sprinting speed into his upward launch. He has only a small space for his approach run and must turn around in order to take off with his back facing the bar. The pole vaulter is able to do much better. He has a long, straight run down the runway and, despite carrying a long pole, the world's best vaulters can achieve speeds of close to 10 metres per second at launch. The elastic fibreglass pole enables them to turn the energy of their horizontal motion $1/2 \, MU^2$ into vertical motion much more efficiently than the high jumper can. Vaulters launch themselves vertically upwards and perform all the impressive gymnastics necessary to curl themselves in an inverted U-shape over the bar while sending their centre of gravity as far below it as possible. Let's see if we can get a rough estimate of how well we might expect them to do. Suppose they manage to transfer all their horizontal running kinetic energy of $1/2 \, MU^2$ first into elastic energy by bending the pole, and then into vertical potential energy of MgH. They will raise their centre of mass a height $H = U^2/2g$.

If the Olympic champion can achieve a 9 m/s launch speed then, since the acceleration due to gravity is $g = 10 \, ms^{-2}$, we expect him to be able to raise his centre of gravity by $H = 4$ metres. If he started standing with his centre of gravity about 1.5 metres above the ground and made it pass 0.5 metres below the bar then he would be expected to clear a bar height of about 1.5 + 4 + 0.5 = 6 metres. In fact, the American champion Tim Mack won the Athens Olympic Gold Medal with a vault of 5.95 metres (19'6" in feet and inches) and had three very close failures at 6 metres, knowing he had already won the Gold Medal, so our very simple estimates turn out to be surprisingly accurate.

17

Superficiality

The periphery is where the future reveals itself.

J.G. Ballard

Boundaries are important. And not only because they keep the wolves out and the sheep in. They determine how much interaction there can be between one thing and another, and how much exposure something local might have to the outside world.

Take a closed loop of string of length p and lay it flat on the table. How much area can it enclose? If you change its shape you notice that by making the string loop longer and narrower you can make the enclosed area smaller and smaller. The largest enclosed area occurs when the string is circular. In that case we know that $p = 2\pi r$ is its perimeter and $A = \pi r^2$ is its area, where r is the radius of the circle of string. So, eliminating r, for *any* closed loop with a perimeter p which encloses an area A, we expect that $p^2 \leq 4\pi A$, with the equality arising only for the case of a circle. Turning things around, this tells us that for a given enclosed area we can make its perimeter as long as we like. The way to do it is to make it increasingly wiggly.

If we move up from lines enclosing areas to surfaces enclosing volumes then we face a similar problem. What shape can maximise the volume contained inside a given surface area? Again, the biggest enclosed volume is achieved by the sphere, with a volume $V = 4\pi r^3/3$ inside a spherical surface of area $A = 4\pi r^2$. So for any closed surface of area A we expect its enclosed volume to obey $A^3 \geq 36\pi V^2$, with equality for the case of the sphere. As before, we see that by making the surface highly crenellated and wiggly we can make the area enclosing a given volume larger and larger. This is a winning strategy that living systems have adapted to exploit.

There are many situations where a large surface area is important. If you want to keep cool, then the larger your surface area the better. Conversely, if you want to keep warm it is best to make it small – this is why groups of newborn birds and animals huddle together into a ball so as to minimise exposed surface. Likewise, the members of a herd of cattle or a shoal of fish seeking to minimise the opportunities for predators will gather themselves into a circular or spherical group to minimise the surface that a predator can attack. If you are a tree that draws moisture and nutrients from the air, then it pays to maximise your surface area interfacing with the atmosphere, so it's good to be branchy with lots of wiggly leafy surfaces. If you are an animal seeking to absorb as much oxygen as possible through your lungs, then this is most effectively done by maximising the amount of tubing that can be fitted into the lung's volume so as to maximise its surface interface with oxygen molecules. And if you simply want to dry yourself after getting out of the shower, then a towel that has lots of surface is best. So towels tend to have a rough pile on their surfaces: they possess much more surface per unit of volume than if they were smooth. This battle to maximise the surface that is enclosing a volume is something we see all over the natural world. It is the reason that 'fractals' so often appear as an evolutionary solution to life's problems. They provide the simplest systematic way to have more surface than you should for your volume.

Is it better for you to stay in one group or to split up into two or more smaller groups? This was a problem faced by naval convoys trying to avoid being found by enemy submarines during the Second World War.

It is better to stay in a big convoy rather than to divide. Suppose that a big convoy covers an area A and the ships are as close together as they can be, so that if we divide the convoy into two smaller ones of area $A/2$ the spacings between ships are the same. The single convoy has a perimeter equal to $p = 2\pi\sqrt{(A/\pi)}$, but the total perimeter of the two smaller convoys equals $p \times \sqrt{2}$, which is bigger. So, the total perimeter distance that has to be patrolled by destroyers to protect the two smaller convoys from being penetrated by submarines is greater than that to be patrolled if it stays as a single convoy. Also, when the submarine searches for convoys to attack, its chance of seeing them is proportional to their diameter, because this is what you see in the periscope. The diameter of the single circular convoy of area A is just $2\sqrt{(A/\pi)}$, whereas the sum of the two diameters of the convoys of area $A/2$ that don't overlap in the field of view is bigger by a factor of $\sqrt{2}$, and so the divided convoy is 41% more likely to be detected by the attacking submarine than is the single one.

18

VAT in Eternity

In this world nothing is certain but death and taxes.

Benjamin Franklin

If you live in the United Kingdom you will know that the sales
tax added on to many purchases is called 'Value Added Tax' or
VAT. In continental Europe it is often called IVA. In the UK it
amounts to an additional 17.5% on the price of a range of goods
and services and is the government's biggest source of tax revenue.
If we suppose that the 17.5% rate of VAT was devised to allow it
to be easily calculated by mental arithmetic what do you expect
the next increase in the rate of VAT to be? And what will the VAT
rate be in the infinite future?

The current value of the VAT rate sounds arcane – why pick
17.5%? But if you have to prepare quarterly VAT accounts for a
small business you ought soon to recognise the niceness of this
funny number. It allows for very simple mental calculation because
17.5% = 10% + 5% + 2.5%, so you know what 10% is immediately (just shift the decimal point one place to the left), then just
halve the 10%, and then halve the 5% that remains and add the
three numbers together. Hence, for example, the VAT on an £80
purchase is just £8 + £4 + £2 = £14.

If the same convenient 'halving' structure for mental arithmetic is maintained, then the next VAT increase will be by an
amount equal to half of 2.5%, or 1.25%, giving a new total

of 18.75% and the new VAT on £80 would be £8 + £4 + £2 + £1 = £15.

We would then have a tax rate that looked like 10% + 5% + 2.5% + 1.25%. To the mathematician this looks like the beginnings of a never-ending series in which the next term in the sum is always half of its predecessor. The current VAT rate is just

$$10\% \times (1 + \tfrac{1}{2} + \tfrac{1}{4})$$

If we continue this series forever we can predict that the VAT rate in the infinite future will be equal to

$$10\% \times (1 + \tfrac{1}{2} + \tfrac{1}{4} + \tfrac{1}{8} + \tfrac{1}{16} + \tfrac{1}{32} + \ldots) = 10\% \times 2$$

where, as we see from page 252, the same series except for the first term is shown to sum to 1, so the sum of the infinite series in brackets is 2. The VAT rate after infinite time is therefore expected to be 20% !

19

Living in a Simulation

Nothing is real.

The Beatles, 'Strawberry Fields Forever'

Is cosmology on a slippery slope towards science fiction? New satellite observations of the cosmic microwave background radiation, the echo of the Big Bang, have backed most physicists' favourite theory of how the Universe developed. This may not be entirely good news.

The favoured model contains many apparent 'coincidences' that allow the Universe to support complexity and life. If we were to consider the 'multiverse' of all possible universes, then ours is special in many ways. Modern quantum physics even provides ways in which these possible universes that make up the multiverse of all possibilities can actually exist.

Once you take seriously the suggestion that all possible universes can (or do) exist then you also have to deal with another, rather strange consequence. In this infinite array of universes there will exist technical civilisations, far more advanced than ourselves, that have the capability to simulate universes. Instead of merely simulating their weather or the formation of galaxies, as we do, they would be able to go further and study the formation of stars and planetary systems. Then, having added the rules of biochemistry to their astronomical simulations, they would be able to watch the evolution of life and consciousness within their computer simulations (all speeded up to occur on whatever timescale was convenient for them). Just as we watch the life cycles of fruit flies, they would be able to follow the evolution of life, watch civilisations grow and communicate with each other, even watch them argue about whether there existed a Great Programmer in the Sky who created their Universe and who could intervene at will in defiance of the laws of Nature they habitually observed.

Within these universes, self-conscious entities can emerge and communicate with one another. Once that capability is achieved, fake universes will proliferate and will soon greatly outnumber the real ones. The simulators determine the laws that govern these artificial worlds; they can engineer fine-tunings that help the evolution of the forms of life they like. And so we end up with a scenario where, statistically, we are more likely to be in a simulated reality than a real one because there are far more simulated realities than real ones.

The physicist Paul Davies has recently suggested that this high probability of our living in a simulated reality is a *reductio ad absurdum* for the whole idea of a multiverse of all possibilities. But, faced with this scenario, is there any way to find out the truth? There may be, if we look closely enough.

For a start, the simulators will have been tempted to avoid the complexity of using a consistent set of laws of Nature in their worlds when they can simply patch in 'realistic' effects. When the Disney company makes a film that features the reflection of light from the surface of a lake, it does not use the laws of quantum electrodynamics and optics to compute the light scattering. That would require a stupendous amount of computing power and detail. Instead, the simulation of the light scattering is replaced by plausible rules of thumb that are much briefer than the real thing but give a realistic looking result – as long as no one looks too closely. There would be an economic and practical imperative for simulated realities to stay that way if they were purely for entertainment. But such limitations to the complexity of the simulation's programming would presumably cause occasional tell-tale problems – and perhaps they would even be visible from within.

Even if the simulators were scrupulous about simulating the laws of Nature, there would be limits to what they could do. Assuming the simulators, or at least the early generations of them, have a very advanced knowledge of the laws of Nature, it's likely they would still have incomplete knowledge of them (some philosophies of science would argue this must always be the case). They may know a lot about the physics and programming needed to simulate a universe, but there will be gaps or, worse still, errors in their knowledge of the laws of Nature. They would, of course, be subtle and far from obvious to us, otherwise our 'advanced' civilisation wouldn't be advanced. These lacunae do not prevent simulations being created and running smoothly for long periods of time, but gradually the little flaws will begin to build up.

Eventually, their effects would snowball and these realities would cease to compute. The only escape is if their creators intervene to patch up the problems one by one as they arise. This is a solution that will be very familiar to the owner of any home computer who receives regular updates in order to protect it against new assaults by viruses or to repair gaps that its original creators had not foreseen. The creators of a simulation could offer this type of temporary protection, updating the working laws of Nature to include extra things they had learned since the simulation was initiated.

In this kind of situation, logical contradictions will inevitably arise and the laws in the simulations will appear to break down occasionally. The inhabitants of the simulation – especially the simulated scientists – will occasionally be puzzled by the observations they make. The simulated astronomers might, for instance, make observations that show that their so-called constants of Nature are very slowly changing.

It's likely there could even be sudden glitches in the laws that govern these simulated realities. That's because the simulators would most likely use a technique that has been found effective in all other simulations of complex systems: the use of error-correcting codes to put things back on track.

Take our genetic code, for example. If it were left to its own devices we would not last very long. Errors would accumulate and death and mutation would quickly ensue. We are protected from this by the existence of a mechanism for error correction that identifies and corrects mistakes in genetic coding. Many of our complex computer systems possess the same type of internal immune system to guard against error accumulation.

If the simulators used error-correcting computer codes to guard against the fallibility of their simulations as a whole (as well as simulating them on a smaller scale in our genetic code), then every so often a correction would take place to the state or the laws governing the simulation. Mysterious changes would occur that would appear to contravene the very laws of Nature that the simulated scientists were in the habit of observing and predicting.

So it seems enticing to conclude that, if we live in a simulated reality, we should expect to come across occasional 'glitches' or experimental results that we can't repeat or even very slow drifts in the supposed constants and laws of Nature that we can't explain.

20

Emergence

A politician needs the ability to foretell what is going to happen tomorrow, next week, next month, and next year. And to have the ability afterwards to explain why it didn't happen.

Winston Churchill

One of the buzz words in the sciences that study complicated things is 'emergence'. As you build up a complex situation step by step, it appears that thresholds of complexity can be reached that herald the appearance of new structures and new types of behaviour which were not present in the building blocks of that complexity. The world-wide web or the stock market or human consciousness seem to be phenomena of this sort. They exhibit collective behaviour which is more than the sum of their parts. If you reduce them to their elementary components, then the essence of the complex behaviour disappears. Such phenomena are common in physics too. A collective property of a liquid, like viscosity, which describes its resistance to flowing, emerges when a large number of molecules combine. It is real but you won't find a little bit of viscosity on each atom of hydrogen and oxygen in your cup of tea.

Emergence is itself a complex, and occasionally controversial, subject. Philosophers and scientists attempt to define and distinguish between different types of emergence, while a few even dispute whether it really exists. One of the problems is that the most interesting scientific examples, like consciousness or 'life', are not understood and so there is an unfortunate extra layer of uncertainty attached to the cases used as exemplars. Here, mathematics can help. It gives rise to many interesting emergent structures that are well defined and suggest ways in which to create whole families of new examples.

Take finite collections of positive numbers like [1,2,3,6,7,9]. Then, no matter how large they are, they will not possess the properties that 'emerge' when a collection of numbers becomes infinite. As Georg Cantor first showed clearly in the nineteenth century, infinite collections of numbers possess properties not shared by any finite subset of them, no matter how large they are. Infinity is *not* just a big number. Add one to it and it stays the same; subtract infinity from it and it stays the same. The whole is not only bigger than its parts, it also possesses qualitatively different 'emergent' features from any of its parts.

A Möbius strip

Many other examples can be found in topology, where the overall structure of an object can be strikingly different from its local structure. The most familiar is the Möbius strip. We make one by taking a thin rectangular strip of paper and gluing the ends together after adding a single twist to the paper. It is possible to make up that strip of paper by sticking together small rectangles of paper in a patchwork. The creation of the Möbius strip then looks like a type of emergent structure. All rectangles that were put together to make the strip have two faces. But when the ends are twisted and stuck together the Möbius strip that results has only one face. Again, the whole has a property not shared by its parts.

21

How to Push a Car

There are only two classes of pedestrian in these days of reck-
less motor traffic – the quick and the dead.

Lord Dewar

There was an old joke that asked, 'Why does a Lada car have a
heater on its rear windscreen?' The answer: 'So that your hands
don't get cold while you are pushing it.' But pushing cars presents
an interesting problem. Suppose you have to push your car into
the garage and bring it to a stop before it hits the back wall. How
should you push it and pull it so as to get into the garage and
stopped as quickly as possible?

The answer is that you should accelerate the car by pushing as
hard as you can for half the distance to be covered and then decel-
erate it by pulling as hard as you can for the other half. The car
will begin stationary and finish stationary and take the least possible
time to do so.[2]

This type of problem is an example of an area of mathematics called 'control theory'. Typically you might want to regu-
late or guide some type of movement by applying a force. The
solution for the car-parking problem is an example of what is
called 'bang-bang' control. You have just two responses: push,
then pull. Domestic temperature thermostats often work like
that. When the temperature gets too high they turn on cooling;
when the temperature gets too low they turn on heating. Over

a long period you get temperature changes that zigzag up and down between the two boundaries you have set. This is not always the best way to control a situation. Suppose you want to control your car on the road by using the steering wheel. A robot driver programmed with the bang-bang control approach would let the car run into the left-hand lane line, then correct to head to the right until it crossed the right-hand lane line, and so on, back and forth. You would soon end up being stopped and invited to blow into a plastic tube before being detained in the local police cells if you followed this bang-bang driving strategy. A better approach is to apply corrections that are proportional to the degree of deviation from the medium position. A swing seat is like this. If it is pushed just a little way from the vertical then it will swing back more slowly than if it is given a big push away from the vertical.

Another interesting application of control theory has been to the study of middle- and long-distance running – and presumably it would work in the same way for horse racing as well as human racing. Given that there is only a certain amount of oxygen available to the runner's muscles, and a limit to how much it can be replenished by breathing, what is the best way to run so as to minimise the time taken to complete a given distance? A control theory solution of bang-bang type specifies that for races longer than about 300 metres (which we know is where anaerobic exercise is beginning and oxygen debt begins) you should first apply maximum acceleration for a short period, then run at constant speed before decelerating at the end for the same short period that you accelerated for initially. Of course, while this may tell you how best to run a time-trial, it is not necessarily the best way to win a race where runners are competing against you. If you have a fast finish or have trained to cope with very severe changes of pace you may well have an advantage over others by adopting different tactics. It will be a brave competitor who sticks to the optimal solution while others are racing into a long lead. Sitting in behind the optimal strategist, sheltering from the wind and getting a free ride before sprinting to victory in the final straight is a very good alternative plan.

22

Positive Feedback

Accentuate the positive. Eliminate the negative. Latch on to
the affirmative. Don't mess with Mr In-between.

Johnny Mercer and Harold Arlen, 'Ac-cent-tchu-ate the Positive'

Earlier this year I had an odd experience while staying in a new
hotel in Liverpool during an unexpected snowfall. The hotel was
a new 'boutique' style of establishment in greatly altered prem-
ises dating back to the nineteenth-century commercial heyday of
the city. In the early morning I had had a tortuous journey through
heavy snow from Manchester, by a slow train that was eventually
halted for a significant time after an announcement by the driver
that the signal cabling along the next stretch of track had been
stolen overnight – another market pointer to the steadily growing
value of copper, I noted. Eventually, by means of coordinated
mobile phone calls, the train moved slowly past all the signals set
by default at red and trundled safely into Lime Street Station.

My hotel room was cold and the temperature outside the snow-
covered skylight windows comfortably below zero. The heating
was underfloor and so rather slow to respond, and it was difficult
to determine whether it was answering to changing the thermostat. Despite assurances from the staff that the temperature would
soon rise, it seemed to get colder and eventually a fan heater was
brought to help out. Someone at reception suggested that because
the heating was new it was important not to turn it up too high.

Much later in the afternoon the building engineer called by, much concerned by the 'old wives' tale' about not turning the heating up too much because it was newly installed and, like me, confused by the fact that the heating was working just fine in the corridors outside – so much so that I left my door open. Fortunately, the master panel for all the room heating was just opposite my door and so we looked together at what it revealed before the engineer investigated the temperature in the room next door where the guest had left for the day. It was very warm next door.

Suddenly, the engineer realised what was the cause of our problem. The heating in the room next door had been wired to the thermostat in my room and my heating to next door's thermostat. The result was a beautiful example of what engineers would call a 'thermal instability'. When my neighbours felt too warm they turned their thermostat down. As a result my room got colder, so I turned the temperature up on my thermostat, which made them feel warmer still, so they turned their thermostat down even further, making me colder still, so I turned my heat up even more . . . Fortunately, they gave up and went out.

This type of instability feeds upon the isolated self-interest of two players. Far more serious environmental problems can arise because of the same sort of problem. If you run lots of fans and air-conditioning to keep cool, you will increase the carbon dioxide levels in your atmosphere, which will retain more of the sun's heat around the Earth, which will increase your demand for cooling. But this problem can't be solved by a simple piece of rewiring.

23

The Drunkard's Walk

Show me the way to go home.
I'm tired and I want to go to bed.
I had a little drink about an hour ago
And it went right to my head.

Irving King

One of the tests of sobriety that is employed by police forces all over the world is the ability to walk in a straight line. Under normal circumstances it is a (literally) straightforward task for an able-bodied person to perform. And if you know how long your stride length is, then you will know exactly how far you will have walked after any number of steps. If your stride is one metre long then after S steps you will have gone S metres from your starting point. But, imagine that, for one reason or another, you are unable to walk in a straight line. In fact, let's suppose you really don't know what you are doing at all. Before you take your next step, choose its direction at random so that you have an equal chance of picking any direction around you and then step one metre in that direction. Now pick a new direction at random and step in that direction. Keep on picking the direction of your next step in this way and you will find that your path wiggles about in a rather unpredictable fashion that has been called the Drunkard's Walk.

An interesting question to ask about the Drunkard's Walk is

how far it will have gone when measured in a straight line from the starting point after the drunkard has taken S steps. We know that it takes the sober walker S one-metre steps to go a straight-line distance S, but it will typically take the drunkard S^2 steps[3] to reach the same distance from the starting point. So, 100 sober steps will get you a distance of 100 metres as the crow flies but the drunkard will usually need 10,000 steps to achieve the same distance.

Fortunately, there is rather more to this numerical insight than the example of the inebriated stroller suggests. The picture of a succession of steps in random directions is an excellent model of what is going on when molecules spread from one place to another because they are hotter than those nearby, and so move faster on the average. They scatter off other molecules one after another in a random way and spread out from the starting point just like the shambling drunkard, taking about N^2 scatterings to go a distance equal to N of the distances that the molecules fly in between scatterings. This is why it takes a significant time to feel the effects of turning on a radiator across a room. The more energetic ('hotter') molecules are 'drunkenly' meandering across the room, whereas the sound waves from the clunking of the air locks in the pipes fly straight across at the speed of sound.

24

Faking It

Blackadder: It's the same plan that we used last time, and the seventeen times before that.

Melchett: E-E-Exactly! And that is what's so brilliant about it! We will catch the watchful Hun totally off guard! Doing precisely what we have done eighteen times before is exactly the last thing they'll expect us to do this time!

Richard Curtis and Ben Elton

One of the most interesting general statistical misconceptions is what a random distribution is like. Suppose that you had to tell whether a particular sequence of events was random or not. You might hope to tell if it was not random by finding a pattern or some other predictable feature. Let's try to invent some lists of 'heads' (H) and 'tails' (T) that are supposed to be the results of a sequence of tosses of a coin so that no one would be able to distinguish them from real coin tossings. Here are three possible fake sequences of 32 coin tosses:

THHTHTHTHTHTHTHTHTHTTTHTHTHTHTHTHH
THHTHTHTHHTHTHTHHHTTHHTHTTHHHTHTTT
HTHHTHTTTHTHTHTHTHHHTHTTTHHHTHTHTHTT

Do they look right? Would you regard them as likely to be real random sequences of 'heads' and 'tails' taken from true coin tosses, or are they merely poor fakes? For comparison, here are three more sequences of 'heads' and 'tails' to choose from:

THHHHTTTTHTTHHHHTTHTHHTTHTTHTHHH
HTTTTHHHTHTTHHHHTTTHTTTTHHTTTTTH
TTHTTHHTHTTTTTHTTHHTTHTTTTTTTHH

If you asked the average person whether these second three lists were real random sequences, most would probably say no. The first three sequences looked much more like their idea of being random. There is much more alternation between heads and tails, and they don't have the long runs of heads and of tails that each of the second trio displays. If you just used your computer keyboard to type a 'random' string of Hs and Ts, you would tend to alternate a lot and avoid long strings, otherwise it 'feels' as if you are deliberately adding a correlated pattern.

Surprisingly, it is the second set of three sequences that are the results of a true random process. The first three, with their staccato patterns and absence of long runs of heads or tails, are the fakes. We just don't think that random sequences can have all those long runs of heads or tails, but their presence is one of the acid tests for the genuineness of a random sequence of heads and tails. The coin tossing process has no memory. The chance of a head or a tail from a fair toss is $1/2$ each time, regardless of the outcome of the last toss. They are all independent events. Therefore, the chance of a run of r heads or r tails coming up in sequence is just given by the multiplication $1/2 \times 1/2 \times 1/2 \times 1/2 \times \ldots \times 1/2$, r times. This is $1/2^r$. But if we toss our coin N times so that there are N different possible starting points for a run of heads or tails, our chance of a run of length r is increased to $N \times 1/2^r$. A run of length r is going to become likely when $N \times 1/2^r$ is roughly equal to 1 — that is, when $N = 2^r$. This has a very simple meaning. If you look at a list of about N random coin tosses, then you expect to find runs of length r where $N = 2^r$.

All our six sequences were of length $N = 32 = 2^5$ so if they are randomly generated we expect there is a good chance that they will contain a run of 5 heads or tails and they will almost surely contain runs of length 4. For instance, with 32 tosses there are 28 starting points that allow for a run of 5 heads or tails, and on average two runs of each is quite likely. When the number of tosses gets large, we can forget about the difference between the number of tosses and the number of starting points and use $N = 2^r$ as the handy rule of thumb.* The absence of these runs of heads or tails is what should make you suspicious about the first three sequences and happy about the likely randomness of the second three. The lesson we learn is that our intuitions about randomness are biased towards thinking it is a good deal more ordered than it really is. This bias is manifested by our expectation that extremes, like long runs of the same outcome, should not occur – that somehow those runs are orderly because they are creating the same outcome each time.

These results are also interesting to bear in mind when you look at long runs of sporting results between teams that regularly play one another, Army vs Navy, Oxford vs Cambridge, Arsenal vs Spurs, AC Milan vs Inter, Lancashire vs Yorkshire. There are often winning streaks where one team wins for many years in a row, although this is not usually a random effect: the same players form the core of the team for several years and then retire or leave and a new team is created.

---

* The result is easily generalised to deal with random sequences where the equally likely outcomes are more than two (H and T here). For the throws of a fair die the probability of any single outcome is $1/6$ and to get a run of the same outcome r times we would expect to have to throw it about 6^r times; even for small values of r, this is a very large number.

25

The Flaw of Averages

Statistics can be made to prove anything – even the truth.

Nöel Moynihan

Averages are funny things. Ask the statistician who drowned in a lake of average depth 3 centimetres. Yet, they are so familiar and seemingly so straightforward that we trust them completely. But should we? Let's imagine two cricketers. We'll call them, purely hypothetically, Flintoff and Warne. They are playing in a crucial test match that will decide the outcome of the match series. The sponsors have put up big cash prizes for the best bowling and batting performances in the match. Flintoff and Warne don't care about batting performances – except in the sense that they want to make sure there aren't any good ones at all on the opposing side – and are going all out to win the big bowling prize.

In the first innings Flintoff gets some early wickets but is taken off after a long spell of very economical bowling and ends up with figures of 3 wickets for 17 runs, an average of 5.67. Flintoff's side then has to bat, and Warne is on top form, taking a succession of wickets for final figures of 7 for 40, an average of 5.71 runs per wicket taken. But Flintoff has the better (i.e. lower) bowling average in the first innings, 5.67 to 5.71.

In the second innings Flintoff is expensive at first, but then proves to be unplayable for the lower-order batsmen, taking 7 wickets for 110 runs, an average of 15.71 for the second innings.

Warne then bowls at Flintoff's team during the last innings of the match. He is not as successful as in the first innings but still takes 3 wickets for 48 runs, for an average of 16.0. So, Flintoff has the better average bowling performance in the second innings as well, this time by 15.71 to 16.0.

Bowler	1st Innings Figures	1st Innings Average	2nd Innings Figures	2nd Innings Average	Combined Figures	Combined Average
Flintoff	3 for 17	5.67	7 for 110	15.71	10 for 127	12.7
Warne	7 for 40	5.71	3 for 48	16	10 for 88	8.8

Who should win the bowling man-of–the-match prize for the best figures? Flintoff had the better average in the first innings and the better average in the second innings. Surely, there is only one winner? But the sponsor takes a different view and looks at the overall match figures. Over the two innings Flintoff took 10 wickets for 127 runs for an average of 12.7 runs per wicket. Warne, on the other hand, took 10 wickets for 88 runs and an average of 8.8. Warne clearly has the better average and wins the bowling award, despite Flintoff having a superior average in the first innings and in the second innings!

All sorts of similar examples spring to mind. Imagine two schools being rated by the average GCSE score per student. One school could have an average score higher than another school on every single subject when they are compared one by one, but then have a lower average score than the second school when all scores were averaged over together. The first school could correctly tell parents that it was superior to the other school in every subject, but the other school could (also legitimately) tell parents that its pupils scored more highly on the average than those at the first school.

There is truly something funny about averages. *Caveat emptor.*

26

The Origami of the Universe

Any universe simple enough to be understood is too simple to
produce a mind able to understand it.

Barrow's Uncertainty Principle

If you want to win bets against over over-confident teenagers then
challenge them to fold a piece of A4 paper in half more than seven
times. They'll never do it. Doubling and halving are processes that
go so much faster than we imagine. Let's suppose that we take our
sheet of A4 paper and slice it in half, again and again, using a laser
beam so that we don't get caught up with the problems of folding.
After just 30 cuts we are down to 10^{-8} centimetres, close to the size
of a single atom of hydrogen. Carry on halving and after 47 cuts
we are down to 10^{-13} centimetres, the diameter of a single proton
forming the nucleus of an atom of hydrogen. Keep on cutting and
after 114 cuts we reach a remarkable size, about 10^{-33} of a centimetre,
unimaginable in our anthropocentric metric units, but not so hard
to imagine when we think of it as cutting the paper in half just
114 times, a lot to be sure, but far from unimaginable. What is so
remarkable about this scale is that for physicists it marks the scale
at which the very notions of space and time start to dissolve. We
have no theories of physics, no descriptions of space, time and
matter that are able to tell us what happens to that fragment of
paper when it is cut in half just 114 times. It is likely that space
as we know it ceases to exist and is replaced by some form of

chaotic quantum 'foam', where gravity plays a new role in fashioning the forms of energy that can exist.[4] It is the smallest length on which we can presently contemplate physical reality to 'exist'. This tiny scale is the threshold that all the current contenders to be the new 'theory of everything' are pushing towards. Strings, M theory, non-commutative geometry, loop quantum gravity, twistors . . . all are seeking a new way of describing what really happens to our piece of paper when it is cut in half 114 times.

What happens if we double the size of our sheet of A4 paper, going to A3, to A2 and so on? After just 90 doublings we have passed all the stars and the visible galaxies, and reached the edge of the entire visible Universe, 14 billion light years away. There are no doubt lots more universe farther away than this, but this is the greatest distance from which light has had time to reach us since the expansion of the Universe began 14 billion years ago. It is our cosmic horizon.

Drawing together the large and the small, we have discovered that just 204 halvings and doublings of a piece of paper take us from the smallest to the largest dimensions of physical reality, from the quantum origins of space to the edge of the visible Universe.

27

Easy and Hard Problems

Finding a hard instance of this famously hard problem can be a hard problem.

Brian Hayes

It takes a long while to complete a large jigsaw puzzle, but just an instant to check that the puzzle is solved. It takes a fraction of a second for your computer to multiply two large numbers together, but it would take you (and your computer) a long time to find the two factors that have been multiplied together to make a big number. It has long been suspected, but never proved or disproved (and there is a one-million-dollar prize for doing either), that there is a real division between 'hard' and 'easy' problems that reflects the amount of calculating time that needs to be used to solve them.

Most of the calculations, or information-gathering tasks, that we have to do by hand, like completing our tax returns, have the feature that the amount of calculating to be done grows in proportion to the number of pieces we have to handle. If we have three sources of income we have to do three times as much work. Similarly, on our computer it takes ten times longer to download a file that is ten times bigger. Ten books will generally take ten times as long to read as one. This pattern is characteristic of 'easy' problems. They may not be easy in the usual sense, but when you add lots of them together the amount of work required doesn't grow very quickly. Computers can easily cope with these problems.

Unfortunately, we often encounter another type of problem that is far less easy to control. Each time we add an extra piece to the calculation, we find that the calculation time required to solve it *doubles*. Very soon the total time required becomes stupendously large, and even the fastest computers on Earth can be easily defeated. These are what we mean by 'hard' problems.[5]

Surprisingly, 'hard' problems are not necessarily horribly complicated or mind-bogglingly difficult. They just involve a lot of possibilities. Multiplying together two large prime numbers is a computationally 'easy' task. You can do it in your head, with pencil and paper or on a calculator, as you wish. But if you give the answer to someone else and ask them to find the two prime numbers that were used in the multiplication, then they might be facing a lifetime of searching with the world's fastest computers.

If you want to try one of these 'hard' problems for yourself, one that sounds deceptively easy, then find the two prime numbers that add up to give 389965026819938.*

These 'trapdoor' operations – so called because, like falling through a trapdoor, it is so much easier to go in one direction than in the reverse – are not altogether bad things. They make life difficult for us but they also make life difficult for people whose lives we are trying to make difficult for a very good reason. All the world's principal security codes exploit trapdoor operations. Every time you shop online or extract cash from an ATM machine you are using them. Your pin number is combined with large prime numbers in such a way that any hacker or computer criminal wanting to steal your account details would have to factor a very large number into the two big prime numbers that were multiplied together to get at it. This is not impossible in principle, but it is impossible in practice in a sensible period of time. A criminal with the world's fastest computer at his disposal might crack

* The answer is 5569 + 389965026814369.

the encryption in several years, but by then the codes and account numbers would have been changed.

For this reason very large prime numbers are very valuable things, and some have been patented when written in a certain form. There is no limit to the number of prime numbers – they go on forever – but there is a largest one that we have been able to check for primeness by ensuring that it has no factors. There is no magic formula that can generate all prime numbers, and it is suspected that no such formula exists. If it did, and it was found, there would be a major crisis in the world. Any government agency that found it would undoubtedly keep it top secret. Any academic who found it and made it public without warning would bring the world tumbling down. All military, diplomatic and banking codes would become easily breakable by fast computers overnight. The world of online commerce would face a serious threat to its continued existence. We would have to move to iris, or fingerprint, or DNA-based recognition systems that relied on unique features of our biochemistry rather than numbers stored in our memories. But these new indicators would still need to be stored in a secure way.

The factoring of prime numbers is a 'hard' problem. Even if it is cracked and shown to be an 'easy' problem by means of a magic formula, you might think that we could just use some other 'hard' problem to encrypt sensitive information so that it still takes ages to reverse the operation and extract it. However, it is known that if one of the problems we believe to be 'hard' could be shown to be 'easy' by means of some new discovery, then that discovery could be used to turn all the other computationally 'hard' problems into 'easy' ones. It really would be a magic bullet.

28

Is This a Record?

I always thought that record would stand until it was broken.

Yogi Berra

Records – remember those little discs of black vinyl that your parents owned that made a noise when spun on a turntable at high speed? Well, mathematicians are more interested in the other sort of records: the biggest, the smallest and the hottest. Are they predictable in any way?

At first you might think not. True, they follow a trend of getting 'better' – they wouldn't be records if they didn't – but how could you predict whether a Michael Johnson or an Ian Thorpe was going to come along and break record after record? Amazingly, the women's world record in the pole vault was broken on eight separate occasions by Yelena Isinbayeva in one year alone. Records like this are not random in a very important sense. Each new record is the result of a competitive effort that is not independent of all the previous attempts at the same feat. Pole vaulters learn new points of technique and continually train to improve their weaknesses and refine their technique. All you can predict about records like this is that they will be set again, eventually, although there might be a very long wait for the next one.

However, there are different sorts of records that arise in sequences of events that are assumed to be independent of each other. Good examples are record monthly rainfalls, record high or low temperatures in one place over hundreds of years, or the heights of the record highest tides. The assumption that each event is independent of its predecessors is a very powerful one that allows us to make a striking prediction about how likely it is that there will be a record – irrespective of what the record is for. It could be rain, snow, fall of leaves, water levels, wind speeds or temperature.

Let's pick on the annual rainfall in the UK as our example. In the first year that we keep records the rainfall must be a record. In year two, if the rainfall is independent of what it was in year 1, then it has a chance of $1/2$ of being a record by beating the year-1 rainfall and a chance of $1/2$ of not beating the year-1 rainfall. So the number of record years we expect in the first two years is $1 + \frac{1}{2}$. In year 3 there are just two ways in which the six possible rankings (i.e. a 1 in 3 chance) of the rainfall in years 1, 2 and 3 could produce a record in year 3. So the expected number of record years after 3 years of record keeping is $1 + \frac{1}{2} + \frac{1}{3}$. If you keep on going, applying the same reasoning to each new year, you will find that after n independent years of gathering data, the expected number of record years is the sum of a series of n fractions:

$$1 + \frac{1}{2} + \frac{1}{3} + \frac{1}{4} + \ldots + 1/n$$

This is a famous series that mathematicians call the 'harmonic' series. Let's label as H(n) its sum after n terms are totalled; so we see that $H(1) = 1$, $H(2) = 1.5$, $H(3) = 1.833$, $H(4) = 2.083$, and so on. The most interesting thing about the sum of this series is that it grows so very slowly as the number of terms increases,* so $H(256) = 6.12$ but $H(1,000)$ is only 7.49 and $H(1,000,000) = 14.39$.

What does this tell us? Suppose that we were to apply our

---

* In fact, when n gets very large H(n) increases only as fast as the logarithm of n and is very well approximated by $0.58 + \ln(n)$.

formula to the rainfall records for some place in the UK from 1748 to 2004 – a period of 256 years. Then we predict that we should find only H(256) = 6.12, or about 6 record years of high (or low) rainfall. If we look at the rainfall records kept by Kew Gardens for this period then this is the number of record years there have been. We would have to wait for more than a thousand years to have a good chance of finding even 8 record years. Records are very rare if events occur at random.

In the recent past there has been growing concern around the world about the evidence for systematic changes in climate, so called 'global warming', and we have noticed an uncomfortably large number of local climatic records in different places. If new records become far commoner than the harmonic series predicts, then this is telling us that annual climatic events are no longer independent annual events but are beginning to form part of a systematic non-random trend.

29

A Do-It-Yourself Lottery

The Strong Law of Small Numbers: There are not enough small numbers to satisfy all the demands placed upon them.

Richard Guy

If you are in need of a simple but thought-provoking parlour game to keep guests amused for a while, then one you might like to try is something that I call the Do-It-Yourself Lottery. You ask everyone to pick a positive whole number, and write it on a card along with their name. The aim is to pick the smallest number *that is not chosen by anyone else*. Is there a winning strategy? You might think you should go for the smallest numbers, like 1 or 2. But won't other people think the same, and so you won't end up with a number that is not chosen by someone else. Pick a very large number – and there are an infinite number of them to choose from – and you will surely lose. It's just too easy for someone else to pick a smaller number. This suggests that the best numbers are somewhere in between. But where? What about 7 or 11? Surely no one else will think of picking 7?

I don't know if there is a winning strategy, but what the game picks up on is our reluctance to think of ourselves as 'typical'. We are tempted to think that we could pick a low number for some reason that no one else will think of. Of course, the reason why opinion polls can predict how we will vote, what we will buy, where we will go on holiday, and how we will respond to an increase in interest rates is precisely because we are all so similar.

I have another suspicion about this game. Although there is an infinite collection of numbers to choose from, we forget about most of them. We set a horizon somewhere around 20, or about twice the number of people in the game, if it is larger, and don't think anyone will pick anything bigger than this. We then exclude the first few numbers up to about 5 on the grounds that they are too obvious for no one else to choose as well and choose from those that remain with roughly equal probability.

A systematic study of preferences would involve playing this game many times over with a large sample of players (say 100 in each trial) to look at the pattern of numbers chosen and the winning choices. It would also be interesting to see how players changed their strategy if they played the game over and over again. A computer simulation of the game is not necessarily of any use because it needs to be told a strategy to adopt. Clearly the numbers are not chosen at random (in which case all numbers would get chosen with equal likelihood). Psychology is important. You try to imagine what others will choose. But the temptation to think that you don't think like anyone else is so strong that almost all of us fall for it. Of course, if there really was a definite strategy for picking the lowest number, everyone would be acting logically to adopt it but that would prevent them choosing a number that no one else chose, and they could never win with that strategy.

30

I Do Not Believe It!

It is quite a three-pipe problem and I beg that you won't speak to me for fifty minutes.

Sherlock Holmes

You are appearing live in a TV game show. The manic presenter shows you three boxes, labelled A, B and C. One of them contains a cheque made out to you for £1 million. The other two contain photos of the presenter. He knows which box contains the cheque, and that cheque will be yours if you pick the right box. You go for Box A. The presenter reaches for Box C and shows everyone that it contains one of the pictures of himself. The cheque must be in Box A or Box B. You picked Box A. The presenter now asks you if you want to stick with your original choice of Box A or switch to Box B. What should you do? Maybe you have an impulse that urges you to switch your choice to Box B, while another voice is saying, Stick with Box A; he's just trying to make you switch to a less expensive option for his employers.' Or perhaps a more rational voice is telling you that it can't possibly make any difference because the cheque is still where it always was and you either guessed it right first time or you didn't.

The answer is remarkable. You should switch your choice to Box B! If you do this you will *double* your chance of picking the box that contains the cheque. Stick with the choice of Box A and you have a 1/3 chance of winning the cheque; switch to Box B and the chance increases to 2/3.

How can this be? At first, there is a 1 in 3 chance that the cheque is in any box. That means a 1/3 chance that it's in A and a 2/3 chance that it's in B or C. When the presenter intervenes and picks a box, it doesn't change these odds because he always picks a box that doesn't contain the cheque. So after he opens Box C there is still a 1/3 chance that the cheque is in A but now there is a 2/3 chance that it is in B because it definitely isn't in C. You should switch.

Still not convinced? Look at it another way. After the presenter opens Box C you have two options. You can stick with your choice of Box A and this will ensure you will win if your original choice was right. Or you can switch boxes, to B, in which case you will be a winner only if your original choice was wrong. Your first choice of Box A will be right 1/3 of the time and wrong 2/3 of the time. So changing your box will get the cheque 2/3 of the time while staying with your first choice will be a winning strategy only 1/3 of the time.

You should by now be convinced by this mind-changing experience.

31

Flash Fires

I will show you fear in a handful of dust

T.S. Eliot, *The Waste Land*

One of the lessons that we have learned from a number of cata-strophic fires is that dust is lethal. A small fire in an old warehouse can be fanned into an explosive inferno by efforts to extinguish it if those efforts blow large amounts of dust into the air where it combusts and spreads the fire through the air in a flash. Anywhere dark and unvisited – under escalators or tiers of seats, or in neglected storage depositories – where dust can build up un-noticed in significant quantities is a huge fire risk.

Why is this? We don't normally think of dust as being partic-ularly inflammable stuff. What is it that transforms it into such a deadly presence? The answer is a matter of geometry. Start with a square of material and cut it up into 16 separate smaller squares. If the original square was of size 4 cm × 4 cm, then each of the 16 smaller squares will be of size 1 cm × 1 cm. The total surface area of material is the same – 16 square centimetres. Nothing has been lost. However, the big change is in the length of the exposed edges. The original square had a perimeter of length 16 cm, but the separate smaller squares each have a perimeter of 4 cm and there are 16 of them, so the total perimeter has grown four times bigger, to 4 × 16 cm = 64 cm.

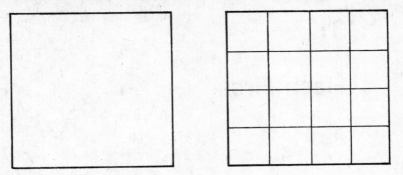

If we do this with a cube, then it would have 6 faces (like a die) of size 4 cm × 4 cm, and each would have an area of 16 sq cm, so the total surface area of the big cube would be 6 × 16 sq cm = 96 sq cm. But if we chopped up the big cube into 64 separate little cubes, each of size 1 cm × 1 cm × 1 cm, the total volume of material would stay the same, but the total surface area of all the little cubes (each with six faces of area 1 cm × 1 cm would have grown to be 64 × 6 × 1 sq cm = 384 sq cm.

What these simple examples show is that if something breaks up into small pieces then the total surface that the fragments possess grows enormously as they get smaller. Fire feeds on surfaces because this is where combustible material can make contact with the oxygen in the air that it needs to sustain itself. That's why we tear up pieces of paper when we are setting a camp fire. A single block of material burns fairly slowly because so little of it is in direct contact with the surrounding air and it is at that boundary with the air that the combustion happens. If it crumbles into a dust of fragments, then there is vastly more surface area of material in contact with the air and combustion occurs everywhere, spreading quickly from one dust fragment to another. The result can be a flash fire or a catastrophic firestorm when the density of dust motes in the air is so great that all the air is caught up in a self-sustaining inferno.

In general, many small things are a bigger fire hazard than one large thing of the same volume and material composition. Careless logging in the forest, which takes all the biggest trees and leaves acres of splintered debris and sawdust all around the forest floor is a topical present-day example.

Powders are dangerous in large quantities. A major disaster happened in Britain in the 1980s when a large Midlands factory making custard powder caught fire. Just a small sprinkling of powdered milk, flour or sawdust over a small flame will produce a dramatic flame bursting several metres into the air. (*Don't try it! Just watch the film.**)

---

* For a sequence of pictures of a demonstration by Neil Dixon for his school chemistry class see http://observer.guardian.co.uk/flash/page/0,,1927850,00.html

32

The Secretary Problem

The chief cause of problems is solutions.

Sevareid's Rule

There is a classic problem about how to make a choice from a large number of candidates; perhaps a manager is faced with 500 applicants for the post of company secretary, or a king must choose a wife from all the young women in his kingdom, or a university must choose the best student to admit from a long list of applicants. When the number of candidates is moderate you can interview them all, weigh each against the others, re-interview any you are unsure about and pick the best one for the job. If the number of applicants is huge this may not be a practical proposition. You could just pick one at random, but if there are N applicants the chance of picking the best one at random is only $1/N$, and with N large this is a very small chance – less than 1% when there are more than 100 applicants. As a route to the best candidate, the first method of interviewing everyone was time-consuming but reliable; the random alternative was quick but quite unreliable. Is there a 'best' method, somewhere in between these two extremes, which gives a pretty good chance of finding the best candidate without spending exorbitant amounts of time in so doing?

There is, and its simplicity and relative effectiveness are doubly surprising, so let's set out the ground rules. We have N known applicants for a 'job', and we are going to consider them in some random order. Once we have considered a candidate we can mark how they stand relative to all the others that we have seen, although we are only interested in knowing the best candidate we have seen so far. Once we have considered a candidate we cannot recall them for reconsideration. We only get credit for appointing the best candidate. All other choices are failures. So, after we have interviewed any candidate all we need to note is who is the best of all the candidates (including them) that we have seen so far. How many of the N candidates do we need to see in order to have the best chance of picking the strongest candidate and what strategy should we adopt?

Our strategy is going to be to interview the first C of the N candidates on the list and then choose the next one of the remaining candidates we see who is better than all of them. But how should we choose the number C? That is the question.

Imagine we have three candidates 1, 2 and 3, where 3 is actually better than 2, who is better than 1; then the six possible orders that we could see them in are

$$123 \qquad 132 \qquad 213 \qquad 231 \qquad 312 \qquad 321$$

If we decided that we would always take the first candidate we saw, then this strategy would pick the best one (number 3) in only two of the six interview patterns so we would pick the best person with a probability of 2/6, or 1/3. If we always let the first candidate go and picked the next one we saw who had a higher rating, then we would get the best candidate in the second (132), third (213), and the fourth cases (231) only, so the chance of getting the best candidate is now 3/6, or 1/2. If we let two candidates go and picked the third one we saw with a higher rating then we would get the best candidate only in the first (123) and third (213) cases, and the chance of getting the best one is again only 1/3. So, when there are three candidates the strategy of letting one go and picking the next with a better rating gives the best chance of getting the best candidate.

This type of analysis can be extended to the situation where the number of candidates, N, is larger than three. With 4 candidates, there are 24 different orderings in which we could see them all. It turns out that the strategy of letting one candidate go by and then taking the next one that is better still gives the best chance of finding the best candidates, and it does so with a chance of success* equal to 11/24. The argument can be continued for any number of applicants and the result of seeing the first 1, or 2, or 3, or 4, and so on, candidates and then taking the next one that is better in order to see how the chance of getting the best candidate changes.

As the number of candidates increases the strategy and outcome get closer and closer to one that is always optimal. Consider the case where we have 100 candidates. The optimal strategy[6] is to see 37 of them and then pick the next one that we see who is better than any of them and then see no one else. This will result in us picking the best candidate for the job with a probability of about 37.1% – quite good compared with the 1% chance if we had picked at random.[7]

Should you use this type of strategy in practice? It is all very well to say that when you are interviewing job candidates you should interview all of them, but what if you apply the same reasoning to a search process for new executives, or your 'search' for a wife, for the outline of the next bestseller or the perfect place to live? You can't search for your whole lifetime. When should you call a halt and decide? Less crucially, if you are looking for a motel to stay in or a restaurant to eat at, or searching for the best holiday deal on line or the cheapest petrol station, how many options

---

* Picking the first candidate or the last always gives a chance of 1/4, letting 2 candidates go gives a chance of 5/12. Letting one go gives a chance of 11/24, which is optimal.

should you look at before you take a decision? These are all sequential choice problems of the sort we have been looking at in the search for an optimal strategy. Experience suggests that we do not search for long enough before making a final choice. Psychological pressures, or simple impatience (either our own or that of others), push us into making a choice long before we have seen a critical fraction, 37 per cent, of the options.

33

Fair Divorce Settlements: the Win–Win Solution

Conrad Hilton was very generous to me in the divorce settlement. He gave me 5,000 Gideon Bibles.

Zsa Zsa Gabor

'It's nice to share, Dad,' our three-year old son once remarked as he looked at my ice cream after finishing his own. But sharing is not so simple. If you need to divide something between two or more people what should you aim to do? It is easy to think that all you need is to make a division that *you* think is fair, and for two people this means dividing the asset in half. Unfortunately, although this might work when dividing something that is very simple, like a sum of money, it is not an obvious strategy when the asset to be shared means different things to different people. If we need to divide an area of land between two countries then each might prize something, like water for agriculture or mountains for tourism, differently. Alternatively, the things being divided might involve undesirables – like household chores or queuing.

In the case of a divorce settlement there are many things that might be shared, but each person places a different value on the different parts. One might prize the house most, the other the collection of paintings or the pet dog. Although you, as a possible mediator, have a single view of the value of the different items to be shared, the two parties ascribe different values to the parts of the whole estate. The aim of a mediator must be to arrive at a division that both side are happy with. That need not mean that the halves are 'equal' in any simple numerical sense.

A simple and traditional way to proceed is to ask one partner to specify a division of assets into two parts and then allow the other partner to choose which of the two parts they want. There is an incentive for the person who specifies the division to be scrupulously fair because they may be on the receiving end of any unfairness if their partner chooses the 'better' piece. This method should avoid any envy persisting about the division process (unless the divider knows something about the assets that the other doesn't – for example, that there are oil reserves under one area of the land). Still, there is a potential problem. The two partners may still value the different parts differently so what seems better for one may not seem so from the other's point of view.

Steven Brams, Michael Jones and Christian Klamler have suggested a better way to divide the spoils between two parties that both feel is fair. Each party is asked to tell the arbiter how they would divide the assets equally. If they both make an identical choice then there is no problem and they immediately agree what to do. If they don't agree, then the arbiter has to intervene.

Suppose the assets are put along a line and my choice of a fair division divides the line at A but your choice divides it at B. The fair division then gives me the part to the left of A and gives you the part to the right of B. In between, there is a left-over portion, which the arbiter divides in half and then gives one part to each of us. In this process we have both ended up with more than the 'half' we expected. Both are happy.

We could do a bit better perhaps than Brams and Co. suggest by not having the arbiter simply divide the remainder in half. We could repeat the whole fair division process on that piece, each choosing where we thought it is equally divided, taking our two non-overlapping pieces so that a (now smaller) piece remains, then divide that, and so on, until we are left with a last piece that (by prior agreement) is negligibly small, or until the choices of how to divide the remnant made by each of us become the same.

If there are three or more parties wishing to share the proceeds fairly then the process becomes much more complicated but is in essence the same. The solutions to these problems have been patented by New York University so that they can be employed commercially in cases where disputes have to be resolved and a fair division of assets arrived at. Applications have ranged from the US divorce courts to the Middle East peace process.

34

Many Happy Returns

You're still so young that you think a month is a long time.

Henning Mankell

If you invite lots of people to your birthday party you might be interested to know how many you need so that there is a better than 50% chance of one of them sharing your birthday. Suppose you know nothing about your guests' birthdays ahead of time, then, forgetting about leap years and so assuming 365 days in a year, you will need at least[8] 253 guests to have a better than evens chance of sharing your birthday with one of them. It's much bigger than 365 divided by two because many of the guests are likely to have the same birthdays as guests other than you. This looks like an expensive birthday surprise to cater for.

A more striking parlour game to plan is simply to look for people who share the same birthday with each other, not necessarily with you. How many guests would you need before there is a better than evens chance of two of them sharing the same birthday? If you try this question on people who haven't worked it out in detail, then they will generally overestimate the number of people required by a huge margin. The answer is striking. With just 23 people[*] there is a 50.7 per cent of two of them sharing a birthday, with 22 the chance is 47.6 per cent and with 24 it is 53.8 per cent.[9]

---

[*] Taking leap years into account makes a small change but does not change this number.

Take any two football teams, throw in the referee and among them you have an odds-on chance of the same birthday being shared by two of them. There is a simple connection with the first problem we considered of having a match with *your* birthday, which required 253 guests for a greater than 50 per cent chance. The reason that any pairing requires just 23 people to be present is because there are so many possible pairings of 23 people – in fact, there are* $(23 \times 22)/2 = 253$ pairings.

The American mathematician Paul Halmos found a handy approximation for this problem in a slightly different form. He showed that if we have a gathering of a large number of people, call the number N again, then we need to take a random selection of at least $1.18 \times N^{1/2}$ of them in order to have a greater than evens chance that two of them will share the same birthday. If you put N = 365 then the formula gives 22.544, so we need 23 people.

One of the assumptions built into this analysis is that there is an equal chance of birthdays falling on any day of the year. In practice, this is probably not quite true. Conception may be more likely at holiday times, and planned birth deliveries by Caesarean section will be unlikely to be planned for Christmas Day or New Year's Eve. Successful sportsmen and sportswomen are another interesting case – you might investigate the birthdays of premier league footballers in a match or members of the UK athletics team. Here, I suspect you would find a bias towards birthdays falling in the autumn. The reason is nothing to do with astrological star signs. The British school year starts at the beginning of September, and so children who have birthdays in September, October and November are significantly older than those in the same school year with June, July and August birthdays at stages

---

* There are 23 for the first choice and then 22 for each of them, so 23 × 22, but then divide by 2 because we don't care about the order in which we find a pair (i.e. 'you and me' is the same as 'me and you').

in life when 6 to 9 months makes a big difference in physical strength and speed. The children with autumn birthdays therefore are rather more likely to make it into sports teams and get all the impetus, opportunity and extra coaching needed to propel them along the road to success as young adults. The same may well be true of other activities that call for different types of maturity as well.

There is a range of activities where we have to give our birth dates as part of a security check. Banks, on-line shopping and airline sites make use of birth dates for part of their passenger security checking. We have seen that taking a birthday alone is not a very good idea. The chance of two customers sharing a date is rather high. You can cut the odds by requiring the year as well and adding a password. You can see that it's essentially the same problem, but instead of having 365 dates that might be shared by two people at your party you are asking for the chance that they share a date and a choice of, say, a string of 10 letters for their password. This reduces the odds on random matches dramatically. If a password was eight alphabetical characters long then there would be 26^{10} choices that could be made, and Halmos' formula would lead us to expect that we need to have 1.28×26^5, or about 15,208,161, customers before there was a 50 per cent chance of two having the same. As of July 2007 the population of the world was estimated to be 6,602,224,175, so having a password sequence that uses just 14 letters makes the number of people required for a 50 per cent chance of a random match larger than the population of the world.

35

Tilting at Windmills

The answer, my friend, is blowin' in the wind.

Bob Dylan

If you travel about the UK you increasingly encounter modern-day windmills dotting parts of the countryside like alien space-ships. Their presence is controversial. They are there to counter atmospheric pollution by offering a cleaner source of power, yet they introduce a new form of visual pollution when inappropri-ately located in pristine (but no doubt windswept) countryside or seascapes.

There are some interesting questions to ask about the windmills, or 'wind turbines' as we call them today. Old style windmills had four sails crossed at the centre like an X. Modern windmills look like aircraft propellers and generally have three arms. There are several factors that have led to the three-armed (or Danish style) windmill becoming so common. Three arms are cheaper than four. So why not two arms? Four-armed windmills have an awkward property that makes them less stable than three-armed ones. Wind-mills with four (or any even number) of arms have the property that when one of the arms is in its highest vertical position, extracting maximum power from the wind, the other end of the sail will be pointing vertically downwards and will be shielded from the wind by the windmill support. This leads to stresses across the sail and a tendency for the windmill to wobble, which can be dangerous in strong winds. Three-armed windmills (and indeed windmills with any odd number of arms) don't suffer from this problem. The three arms are each 120 degrees apart, and when any one is vertical neither of the other two arms can be vertical as well. Of course, three arms will catch less wind than four and so will have to rotate faster in order to generate the same amount of power.

The efficiency of windmills is an interesting question, which was first solved back in 1919 by the German engineer Albert Betz. A windmill's sails are like a rotor that sweeps out an area of air, A, which approaches it at a speed U and then passes on at another, lower, speed V. The loss in air speed as a result of the sails' action is what enables power to be extracted from the moving air by the windmill. The average speed of the air at the sails is $1/2 (U + V)$. The mass of air per unit time passing through the rotating arms is $F = DA \times 1/2 (U + V)$, where D is the density of air. The power generated by the windmill is just the difference in the rate of change of kinetic energy of the air before and after passing the rotors. This is $P = 1/2 FU^2 - 1/2 FV^2$. If we use the formula for F in this equation, then we see that the power generated is

$$P = 1/4 \, DA(U^2 - V^2)(U + V)$$

But if the windmill had not been there the total power in the undisturbed wind would have just been $P_0 = 1/2 \, DAU^3$. So, the efficiency of the windmill in extracting power from the moving mass of air is going to be P/P_0, which equals 1, indicating 100 per cent efficiency as a power generator, when $P = P_0$. Dividing the last two formulae, we have

$$P/P_0 = 1/2 \, \{1 - (V/U)^2\} \times \{1 + (V/U)\}$$

This is an interesting formula. When V/U is small, P/P_0 is close to $1/2$; as V/U approaches its maximum value of 1, so P/P_0 tends to zero because no wind speed has been extracted. Somewhere in between, when $V/U = 1/3$, P/P_0 takes its maximum value. In that case, the power efficiency has its maximum value, and it is equal to $16/27$, or about 59.26%. This is Betz's Law for the maximum possible efficiency of a windmill or rotor extracting energy from moving air. The reason that the efficiency is less than 100% is easy to understand. If it had been 100% then all the energy of motion of the incoming wind would have to be removed by the windmill sails – for example, this could be achieved by making the windmill a solid disc so it stopped all the wind – but then the rotors would not turn. The downstream wind speed (our quantity V) would be zero because no more wind would flow through the windmill sails.

In practice, good wind turbines manage to achieve efficiencies of about 40%. By the time the power is converted to usable electricity there are further losses in the bearings and transmission lines, and only about 20% of the available wind power will ultimately get turned into usable energy.

The maximum power that can be extracted from the wind by any sail, rotor, or turbine, occurs when $V/U = 1/3$, so $P_{max} = (8/27)$ $\times D \times A \times U^3$. If the diameter of the rotor circle is d then the area is $A = \pi d^2/4$ and if the windmill operates at about 50% efficiency, then the power output is about $1.0 \times (d/2 \text{ m})^2 \times (U/1 \text{ ms}^{-1})^3$ watts.

36

Verbal Conjuring

> Posh and Becks failed to appear from their hotel room, having
> been confused by the 'Do Not Disturb' sign hanging on the
> inside of the door.
>
> Angus Deayton

Skilful conjuring, viewed close up, can be very perplexing, and it
becomes astonishing if the conjuror shows you how it was done.
How easily he misled you; how blind you were to what was going
on right under your nose; how simple it was. You soon realise how
incompetent a judge you would be in the face of claims of spoon
bending or levitation. Scientists are the easiest to fool: they are
not used to Nature conspiring to trick them. They believe almost
everything they see is true. Magicians believe nothing.

In this spirit, I want to relate to you a little mathematical story,
adapted from a version by Frank Morgan, that is an example of
verbal conjuring. You keep track of everything in the story but
something seems to disappear in the telling – a sum of money no
less – and you have to figure out where it went; or, indeed, whether
it was ever there.

Three travellers arrive at a cheap hotel late at night, each with
just £10 in his wallet. They decide to share one large room and
the hotel charges them £30 for a one-night stay, so they each put
in £10. After they have gone up to their room with the bell boy
carrying their bags, the hotel receptionist receives an email from

the head office of the hotel chain saying that they are running a special offer and the room rate is reduced to £25 for guests staying tonight. The receptionist, being scrupulously honest about such things, immediately sends her bell boy back to the room of the three new guests with a £5 note for their rebate. The bell boy is less scrupulous. He hadn't received a tip for carrying their bags and he can't see how to split £5 into three, so he decides to keep £2 for himself, as a 'tip', and give the three guests a rebate of £1 each. Each of the three guests has therefore spent £9 on the room and the bell boy has £2 in his pocket. That makes a total of £29. But they paid £30 – what happened to the other £1?*

---

* By following the argument closely, you will see that there is no missing pound. In the end the three guests have a total of £3, the bell boy has £2, and the hotel has £25.

37

Financial Investment with Time Travellers

I'm, not a soothsayer! I'm a con man, that's all! I can't make
any genuine predictions. If I *could* have foreseen how this was
going to turn out I'd have stayed at home.

René Goscinny and Albert Uderzo, *Asterix and the Soothsayer*

Imagine that there are advanced civilisations in the Universe that
have perfected the art (and science) of travelling in time. It is
important to realise that travelling into the future is totally uncon-
troversial. It is predicted to occur in Einstein's theories of time
and motion that so accurately describe the world around us and
is routinely observed in physics experiments. If two identical twins
were separated, with one staying here on Earth while the other
went on a round space-trip then the space-travelling twin would
find himself younger than his stay-at-home brother when he
returned to meet him on Earth. The returning twin has time travelled into the future of the stay-at-home twin.

So, time travelling into the future is just a matter of practicality:
can you build means of transportation that can withstand the stresses
and achieve the velocities close to the speed of light necessary to
make it a noticeable reality? Time travel into the past is another
matter entirely. This is the realm of so-called changing-the-past paradoxes, although most of them are based upon a misconception.*

* See J.D. Barrow, *The Infinite Book* (Vintage), for details.

Here is an observational proof that time travellers are not engaged in systematic economic activity in our world. The key economic fact that we notice is that interest rates are not zero. If they are positive then backward time travellers could use their knowledge of share prices gleaned from the future to travel into the past and invest in the stocks that they know will increase in value the most. They would make huge profits everywhere across the investment and futures markets. As a result, interest rates would be driven to zero. Alternatively, if interest rates were negative (so that investments are worth less in the future), time travellers could sell their investment at its current high price and repurchase it in the future at a lower price so as to travel backwards in time to resell at a high price again. Again, the only way for the markets to stop this perpetual money machine is by driving interest rates to zero. Hence, the observations that there exist interest rates that are not zero means this type of share-dealing activity is not being carried out by time travellers from the future.*

The same type of argument would apply to casinos and other forms of gambling. Indeed, this might be a better target for the time travellers because it is tax free! If you know the future winner of the Derby or the next number on which the roulette ball will stop, then a short trip to the past will guarantee a win. The fact that casinos and other forms of gambling still exist – and make a significant profit – is again a telling argument that time-travelling gamblers do not exist.

These examples might seem rather fanciful but, on reflection, might not the same arguments not be levelled against many forms of extrasensory perception or parapsychic forms of knowledge? Anyone who could foresee the future would have such an

---

* This argument was first made by the Californian economist Marc Reinganum. For more conservative investors, note that £1 invested in 2007 in a savings account giving 4% compound interest would have risen in value to £1 × $(1 + 0.04)^{1000}$ = 108 million billion pounds by the year 3007. However, this will probably still be

enormous advantage that they would be able to amass stupendous wealth quickly and easily. They could win the lottery every week. If reliable intuitions about the future did exist in some human (or pre-human) minds, then they would provide the owners with an enormous evolutionary advantage. They would be able to foresee hazards and plan for the future without uncertainty. Any gene that endowed its owners with that insurance policy against all eventualities would spread far and wide and its owners would soon come to dominate the population. The fact that psychic ability is apparently so rare is a very strong argument against its existence.

38

A Thought for Your Pennies

One of the evils of money is that it tempts us to look at it
rather than at the things that it buys.

E.M. Forster

Coins cause delays. Buy something that costs 79 pence (or 79
cents) and you will be rummaging in the depths of your purse
for an exact combination of coins to make up the amount. If
you don't, and offer £1 (or $1) instead, then you are about to
receive even more small change and it will take you even longer
to find the change you need next time. The question is: what is
the collection of different coin values that is best for making up
your change?

In the UK (and Europe) we have coins of value 1, 2, 5, 10, 20
and 50 in order to make up £1 (or 1 Euro). In the USA there are
1, 5, 10, and 25 cent coins with which to make change of $1 or
100 cents. These systems are 'handy' in the sense that the smallest
number of coins needed to make up any amount up to a 100 is
used by following a simple recipe that computer scientists call a
'greedy algorithm'. This recipe, and hence its name, makes up
the total by taking as many of the largest coins as possible first,
then doing the same with the next largest, and so on. If we
wanted to make 76 cents then this could best be done by taking
only three coins: one 50c, one 25c, and one 1c. In the UK, for 76
pence we would need one 50p, one 20p, one 5p and one 1p, so

we require a minimum of four coins – one more than the American system needs despite having more denominations to choose from.

My last year in school is memorable for the fact that it contained the 'D-Day' date of 15 February 1971, when the UK changed to a decimal coinage system. Previously, the old system of £sd, pounds, shillings and pence, had a large number of coins with unusual values (for historical reasons). There were 240 old pennies (labelled d, from the Latin *denarii*) in £1, and in my childhood there were coins with values, ½d, 1d, 3d, 6d, 12d, 24d, and 30d, which had the common names halfpenny, penny, threepenny bit, sixpence, shilling, florin and half crown. This system of coins does not have the 'handy' property that it makes change most efficiently by following the greedy algorithm. If you wanted to make up 48d (4 shillings), then the algorithm would tell you to use one 30d, one 12d and one 6d – three coins in total. But you could have done it more efficiently by using two 24p pieces. In previous centuries there existed a 4d coin, called a groat, which had the same effect on the change making. A greedy algorithm would make 8d with three coins, a 6d and two 1d, but it can be done with just two groats. This situation arises when we double the value of one of our coins that are bigger that 2d (the 24d or the 4d) because there is not another coin with that value. This cannot happen with the modern US, UK and Euro denominations.

All the modern systems that are in use pick from a similar set of round numbers, 1, 2, 5, 10, 20, 25 and 50, for their coins. But is this the best possible set to choose from? The numbers are relatively easy to add up and combine, but are they the optimal when it comes to making change with as few coins as possible?

A few years ago Jeffrey Shallit, based in Waterloo, Ontario, carried out a computer investigation of the average number of coins you need to make up any amount of change from 1 cent to 100 cents in the US system with different choices of coin denominations. If you use the actual collection of 1, 5, 10 and 25 cent denominations, then the average number of coins you will need to make up every amount from 1 to 100 cents turns out to be 4.7. If you only had a 1 cent coin you would need 99 coins to make up 99 cents and the average number for all amounts from 1 to 100 is 49.5 – the worst possible. If you just have a 1 cent and a 10 cent coin then the average falls to 9. The interesting question is whether we can make the average lower with a different set of 4 coins than the actual set of 1, 5, 10 and 25 cent pieces. The answer is 'yes', and there are two different sets of coins that do better. If there were 1, 5, 18 and 29 cent coins or 1, 5, 18 and 25 cent coins only, then the average number of coins needed to make every amount from 1 to 100 cents is only 3.89. The best one to take is the 1, 5, 18 and 25 cent solution as this only requires one small change from the actual – replace the 10 cent dime with a new 18 cent piece!

We can see what happens in the UK or the Euro system when we subject it to a similar analysis and ask what new coin should we add to our currency to make it as efficient as possible. These systems now have coins with 1, 2, 5, 10, 20, 50, 100 and 200 denominations (i.e., in the UK we now have £1 and £2 coins and no £1 bank note). Suppose we work out the average number of coins needed to make up all amounts from 1 (penny or cent) to 500 in these systems. It is 4.6. But this could be reduced to 3.92 by adding a coin of value 133 or 137 pence or cents.

39

Breaking the Law of Averages

There are three types of lies – lies, damned lies, and statistics.

Benjamin Disraeli

Our look at the generation of wind power in Chapter 35 hides an interesting subtlety about statistics that it is good to be aware of. We all know that there are three types of lies – lies, damned lies, and statistics, as Disraeli warned – but it is one thing to know that there is something to beware of, quite another to know where the danger lurks. Let's see how you could be misled by a sneaky use of statistics in the case of wind power. We know that the power available in the wind is proportional to the cube of the wind's speed, V^3. To avoid writing in all the other quantities involved, like the density of the air, which don't change, let's just say that the power generated per unit of time is equal to this, so $P = V^3$. For simplicity we assumed that the average speed of the wind over a year was 5 metres per second and so the power generated over one year would be $5^3 \times 1 \text{ yr} = 125$ units.

In practice, the average speed of the wind is not what the wind speed always is. Suppose that we allow for a very simple variation. Assume that the wind's speed is zero for half of the days of the year and equal to 10 metres per second for the other half. The average speed over the year is still $\frac{1}{2} \times 10 = 5$ metres per second.

But what is the power generated now? For half of the year it must be zero because the wind speed is zero; for the other half it is $\frac{1}{2} \times 10^3 = 500$. The total generated over the year is therefore 500 units, much more than we get if we just take the average speed. The days with the higher wind speeds vastly over compensate in power output for the days with no wind at all. In reality, the distribution of wind speeds over a year is more complicated than the simple one used here, but it has the same property of producing much more power gain when the wind speed is above average than power loss when it is below average This is a situation that contravenes what is proverbially known as the *law of averages*. In fact, this is not a general law at all, but just the intuition that many people have that variations all even out in the end: that is, in the long run there will be as many gains as losses of comparable sizes above and below the average behaviour. This is true only for random variations governed by statistics that have a special symmetric pattern. In our problem of the wind power generation this is not the case, and so the above-average winds have a much bigger effect than the below-average winds. Beware of Mister Average – and Ms Average too.

40

How Long are Things Likely to Survive?

Statistics are like a bikini. What they reveal is suggestive, but what they conceal is vital.

Aaron Levenstein

Statistics are powerful tools. They seem to tell you something for nothing. They often predict the outcomes of our elections on the basis of just a few early samples of voters' decisions in one or two constituencies. To the outsider it appears that they are able to draw conclusions from almost no evidence at all. One of my favourite examples of this apparent wizardry is with regard to predicting the future. If we have some institution or tradition that has been in existence for a given period of time, how long should we expect it to survive in the future? The key idea is rather simple. If you were to see something at a random moment, then there is a 95% chance that you will be glimpsing it during the middle 95% of its total lifetime. Consider an expanse if time equal to 1 time unit and take out the middle interval of 0.95 so there is an interval of length 0.025 at the beginning and the end:

0 ◄— 0.025 —► A ◄— 0.95 —► B ◄— 0.025 —► 1

If you observed at A, then the future occupies 0.95 + 0.025 = 0.975 and the past occupies 0.025. So the future is 975/25 = 39 times longer than the past. Similarly, if you observed at time B then the future would be only a fraction, 1/39, of the past.

When we observe something in history, like the existence of the Great Wall of China or Cambridge University, then, if we assume that we are not observing it at a special time in history, in the sense that there is no reason why we should be particularly near its beginning or its end, we can predict how long we should expect it to exist in the future with a 95% level of confidence.* Our diagram shows us the numbers we need. If an institution has been around for Y number of years and we are observing it at a random time in history, then we can be 95% sure that it will exist for at least Y/39 years but not for more than 39 × Y years.[10] In 2008 Cambridge University has been in existence for 800 years. This formula predicts that its future longevity beyond that date has a 95% chance of being for at least 800/39 = 20.5 years, approximately, and at most 800 × 39 = 31,200 years. The United States was declared an independent country in 1776. It has a 95% chance of lasting more than 5.7 years but less than 8,736 years. The human race has been around for about 250,000 years, so it will survive for between 6,410 and 9,750,000 years with 95% confidence. The Universe has been expanding for 13.7 billion years. If this is how long it has been in existence, then with 95% probability it will last for more than another 351 million years and less than 534.3 billion years.

* It is tempting to continue applying this rule to everything and anything, but a word of caution is in order. Some things have a life expectancy that is determined by something more than chance alone (such as biochemistry). When this probability rule is applied in situations where there is a non-random process determining the timescale for change, it leads to a wrong conclusion about the longest (or the shortest) expected survival time. The formula says there is a 95% chance that a 78-year-old person will live for between 2 and 3042 more years. However, we would expect that the chance of them living for more than 50 years is precisely zero. Age 78 was not a random time in that person's life: it is much closer to the biological endpoint than to the day when they were born. For more detailed discussion see http://arxiv.org/abs/0806.3538.

Try it on some things that have been around for rather less time. My house has been standing for 39 years, so I should be 95% sure it won't succumb to some random disaster in the next year but I (or somebody) should be surprised at the 95% level if it survives for 1,521 years. You can try it out on football club managers, businesses, countries, political parties, fashions and the runs of West End plays. It is interesting to make predictions too. As I write this, Gordon Brown has been Prime Minister of the UK since 27 June 2007 – almost exactly three months. He addressed his first Labour Party Conference as Prime Minister yesterday, but we could tell him that with 95% confidence he will be in office for at least another two or three days but not more than 9¾ years. By the time this book is published you might be able to test this prediction!

41

A President who Preferred the Triangle to the Pentagon

The triangle appears to require no specialist ability to play.

The Triangle, *Wikipedia*

James Garfield was the 20th President of the United States. You probably know nothing about him except perhaps that he was shot by an unhappy member of the public who had wanted to be appointed to a federal post, on 2 July 1881, after just four months in office, and he died ten weeks later. Strangely, the lethal bullet that lodged in his body was never found, despite Alexander Graham Bell being commissioned to invent a metal detector for the purpose. He succeeded in making such a device, but it was not very effective, primarily we suspect because Garfield's bed at the White House had a metal frame – a very rare thing at the time – and nobody suspected that it was the cause of the instrument's malfunction. In retrospect, the real cause of his death was careless medical care that punctured his liver. Thus, he became the second President to be assassinated and served the second shortest term of office. Despite this sorry end, his name lives on in a small way because of an unusual contribution he made to mathematics.

Garfield had originally intended to become a mathematics teacher after he graduated from Williams College in 1856. He taught classics for a while and made an unsuccessful attempt to become a school headmaster, but his patriotism and strong views led him to run for public office, and he was elected to the Ohio State Senate just three years later and qualified as a barrister in 1860. He left the Senate and joined the army in 1861 and rose swiftly up the ranks to become a major-general, before moving sideways into the House of Representatives two years later. There he remained for 17 years until he became the Republican Presidential candidate in 1880, winning a narrow victory in the popular vote over the Democratic party candidate, Winfield Hancock, on a ticket that promised to improve education for all Americans. Garfield is still the only person ever to be elected President directly from the House of Representatives.

Garfield's most interesting contribution was nothing to do with politics at all. In 1876, while serving in the House, he liked to join with fellow members of Congress to talk about subjects of a mind-broadening sort. Garfield presented for his colleagues' amusement a new proof of Pythagoras's theorem for right-angled triangles. Later, he published it in the *New England Journal of Education*, where he remarks that 'we think it something on which the members of both houses can unite without distinction of party'.

Mathematicians had been teaching this theorem to students for more than 2,000 years, and they usually stayed close to the proof given by Euclid in his famous *Elements*, which he wrote in Alexandria in about 300BC. This proof was by no means the first. Both the Babylonians and the Chinese had good proofs, and the ancient Egyptians were well acquainted with the theorem and were able to use it in their adventurous building projects. Of all the proofs of Pythagoras's theorem that have been found over the centuries, Garfield's is one of the simplest and the easiest to understand.

Take a right-angled triangle and label its three sides a, b and c. Make a copy of it and lay the two copies down so they form a V-shaped crevice with a flat base. Now join up the two highest points of the two triangles to make a lop-sided rectangle – a shape that

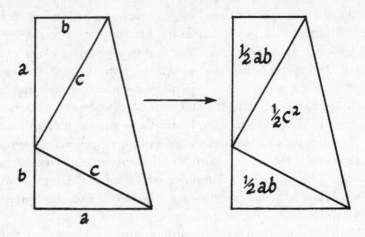

we call a trapezium (or, in America, a trapezoid). It is a four-sided figure, which has two of its sides parallel, shown above.

Garfield's figure consists of three triangles – the two copies of the one we started with and the third one that was created when we drew the line between their two highest points. He now simply asks us to work out the area of the trapezium in two ways. First, as a whole it is the height, a+b, times the average width ½ (a+b); so the area of the trapezium is ½ $(a+b)^2$. To convince yourself of this you could change the shape of the trapezium, turning it into a rectangle by making the two widths equal. The new width would be ½(a+b).

Now, work out the area another way. It is just the sum of the areas of the three right-angled triangles that make it up. The area of a right-angled triangle is just one half of the area of the square you would get by joining two copies of it together along the diagonal, and so it is one-half of its base length times its height. The total area of the three triangles is therefore ½ba + ½c^2 + ½ba = ba + ½c^2, as shown.

Since these two calculations of the total area must give the same total, we have:

$$\tfrac{1}{2}(a+b)^2 = ba + \tfrac{1}{2}c^2.$$

That is,

$$\tfrac{1}{2}(a^2 + b^2 + 2ab) = ba + \tfrac{1}{2}c^2$$

And so, multiplying by two:

$$a^2 + b^2 = c^2$$

just as Pythagoras said it should.

All prospective candidates in the American elections should be asked to give this proof during the televised Presidential debates.

42

Secret Codes in Your Pocket

No one can buy or sell who does not have the mark.

Book of Revelation

Codes mean spies, secret formulae and countries at war – right? Wrong: codes are all around us, on credit cards, cheques, bank notes and even on the cover of this book. Sometimes codes play their traditional role of encrypting messages so that snoopers cannot easily read them or preventing third parties from raiding your online bank account, but there are other uses as well. Databases need to be kept free of innocent corruption as well as malicious invasion. If someone types your credit card number into their machine but gets one digit wrong (typically swopping adjacent digits, like 43 for 34, or getting pairs of the same digit wrong, so 899 becomes 889) then someone else could end up being charged for your purchase. Enter a tax identification number, an airline ticket code or a passport number incorrectly and the error could spread through part of the electronic world, creating mounting confusion.

The commercial world has tried to counter this problem by creating means by which these important numbers can be self-checking to tell their computers whether or not the number being entered qualifies as a valid air ticket or bank note serial number. There is a variety of similar schemes in operation to check the validity of credit card numbers. Most companies use a system introduced by IBM for credit-card numbers of 12 or 16 digits. The process is a bit laborious to carry out by hand, but can be checked in a split second by a machine, which will reject the card number input if the digits fail the test because of error or a naively faked card.

Take an imaginary *Visa* card number

$$4000\ 1234\ 5678\ 9314$$

First, we take every other digit in the number, going from left to right, starting with the first (i.e the odd-numbered positions), and double it to give the numbers 8, 0, 2, 6, 10, 14, 18, 2. Where there are double digits in the resulting numbers (like 10, 14, 18) we add the two digits together to get (1, 5, 9), which has the same effect as subtracting 9. The list of doubled numbers now reads 8, 0, 2, 6, 1, 5, 9, 2. Now we add them together and then add all the numbers in between (i.e. the even-numbered positions) that we missed out the first time (0, 0, 2, 4, 6, 8, 3, 4). This gives the sum, in order, as

$$8+0+0+0+2+2+6+4+1+6+5+8+9+3+2+4 = 60$$

In order for the card number to be valid, this number must be exactly divisible by 10, which in this case it is. Whereas, if the card number was 4000 1234 5678 9010, then the same calculation would have generated the number 53 (because the card number only differs in the last and third-last digits) and this is not exactly divisible by 10. This same procedure works to verify most credit cards.

This checking system will catch a lot of simple typing and reading errors. It detects all single-digit errors and most adjacent swops (although 90 becoming 09 would be missed).

Another check-digit bar code that we are always encountering (and totally ignoring, unless we are a supermarket check-out assistant) is the Universal Product Code, or UPC, which was first used on grocery products in 1973 but has since spread to labelling most of the items in our shops. It is a 12-digit number that is represented by a stack of lines, which a laser scanner can read easily. The UPC has four parts: below the bars, two separate strings of 5 digits are set between two single digits. For example, on the digital camera box on my desk at the moment this looks like

$$0 \; 74101 \; 40140 \; 0$$

The first digit identifies the sort of product being labelled. The digits 0, 1, 6, 7, 9 are used for all sorts of products; digit 2 is reserved for things like cheese, fruit and vegetables that are sold by weight; digit 3 is for drugs and heath products; digit 4 is for items that are to be reduced in price or linked to store loyalty cards, and digit 5 is for items linked to any 'money-off' coupons or similar offers. The next block of five digits identifies the manufacturer – in my case Fuji – and the next five are used by the manufacturer to identify the product by its size, colour and features other than price. The last digit – 0 here – is the check digit. Sometimes it isn't printed but is just represented in the bars for the code reader so it can accept or reject the UPC. The UPC is generated by adding the digits in the odd-numbered positions ($0+4+0+4+1+0 = 9$), multiplying by three ($3 \times 9 = 27$), and then adding the digits in the even-numbered positions to the result ($27+7+1+1+0+4+0 = 40 = 4 \times 10$), and checking that it is divisible by 10 – which it clearly is.

That just leaves the bars. The overall space inside the outermost single digits (our two zeros) is split into seven regions, which are filled with a thickness of black ink, which depends on the digit being represented, with light bands and dark bands alternating. On each end of any UPC there are two parallel 'guard bars' of equal thickness, which define the thickness and separation scale that is used by the lines and spaces in between. There is a similar set of four central bars, two of which sometimes drop below the others, which separate the manufacturer's ID from the product information and carry no other information. The actual positions and thicknesses of the bars form a binary code of 0s and 1s. An odd number of binary digits is used to encode the manufacturer's details, while an even number is used for the product information. This prevents confusion between the two and enables a scanning device to read these numbers from right to left or left to right and always know which block it is looking at. And you thought life was simple.

43

I've Got a Terrible Memory for Names

The 't' is silent, as in Harlow.

Margot Asquith, on her name being mispronounced by Jean Harlow

If you have ever had to make a note of someone's name over the telephone, then you will know that it is a tricky business being sure of the spelling. Usually you respond to uncertainty by asking them to spell out their name. I recall how my doctoral research supervisor, Dennis Sciama, whose unusual surname was pronounced 'Sharma', could spend a significant amount of his working day spelling out his name to phone callers who did not know him.

There are occasions when oral and written messages can't be repeated or have been wrongly written, and you want to minimise the possibility of missing out on a person's real identity when you look them up in your files. The oldest scheme in operation to try to ameliorate this problem is called the Soundex phonetic system, and dates from about 1918, when it was invented by two Americans, Robert Russell and Margaret Odell, although it has gone through various small modifications since then. It was originally designed to help with the integrity of census data that was gathered orally, and was then used by airlines, the police and ticket booking systems.

The idea was to encode names so that simple spelling variants, like Smith and Smyth, or Ericson and Erickson, that sounded the same were coded as the same, so that, if you entered one of the group, then the other variants would appear as well, thus ensuring that you were not missing one in the filing system. Anyone searching for relatives or ancestors, especially immigrants with foreign names that might have been slightly modified, would find this encoding useful. It will automatically seek out many of the close variants that you would otherwise have to search out one by one and will also find variants you hadn't even thought of. Here is how it works for names.

1. Keep the first letter of the name, whatever it is.
2. Elsewhere, delete all occurrences of the following letters: a,e,i,o,u,h,y,w
3. Assign numbers to the other letters that remain so that:
   b,f,p,v all become 1
   c,g,j,k,q,s,x,z all become 2
   d and t both become 3 and l becomes 4
   m and n become 5 and r becomes 6
4. If two or more letters with the same number were next to one another in the original full name keep only the first of them.
5. Finally, record only the first four characters of what you have left. If you have fewer than four then just add zeros on the end to make the string up to a length of four.

My name is John, and the process will change it to Jn (after steps 1 and 2), then J5 (after step 3), and the final record is J500. If your name is Jon you will get the same result. Smith and Smyth both become S530. Ericson, Erickson, Eriksen and Erikson all give the same record of E6225.

44

Calculus Makes You Live Longer

As a math teacher, I understand how important it is for
students to see that mathematics can connect with life.
Mortuary science gives me a novel and unique way to do that.
After all, what could be more universal in life than death?
Once my students learn about rates of decay and embalming
theory, they seem eager to return to the study of calculus with
a renewed rigor.

Professor Sweeney Todman, Mathemortician

The difference between an amateur and a professional is that as
an amateur one is at liberty to study only those things one likes,
but as a professional one must also study what one doesn't like.
Consequently, there are parts of a mathematical education that
will seem laborious to a student, just as all those hours of winter
running in the cold and rain will be unattractive, but essential, to
the aspiring Olympic athlete. If students asked why they needed
to learn some of the more intricate and unexciting parts of calculus,
I used to tell them this story, one that the Russian physicist George
Gamow tells us in his quirky autobiography, *My World Line*. It is
about the remarkable experience of one of Gamow's friends, a
young physicist from Vladivostok called Igor Tamm, who went
on to share the Nobel prize for physics in 1958 for his part in

discovering and understanding what is now known as the 'Cerenkov Effect'.

In the Russian revolutionary period, Tamm was a young professor teaching physics at the University of Odessa in the Ukraine. Food was in short supply in the city, and so he made a trip to a nearby village, which was under the apparent control of the communists, in an attempt to trade some silver spoons for something more edible, like chickens. Suddenly, the village was captured by an anti-communist bandit leader and his militia, armed with rifles and explosives. The bandits were suspicious of Tamm, who was dressed in city clothes, and took him to their leader, who demanded to know who he was and what he did. Tamm tried to explain that he was merely a university professor looking for food.

'What kind of professor?' the bandit leader asked.

'I teach mathematics,' Tamm replied.

'Mathematics?' said the bandit. 'All right! Then give me an estimate of the error one makes by cutting off Maclaurin's series at the nth term.[11] Do this and you will go free. Fail, and you will be shot!'

Tamm was not a little astonished. At gunpoint, somewhat nervously, he managed to work out the answer to the problem – a tricky piece of mathematics that students are taught in their first course of calculus in a university degree course of mathematics. He showed it to the bandit leader, who perused it and declared 'Correct! Go home!'

Tamm never discovered who that strange bandit leader was. He probably ended up in charge of university quality assurance somewhere.

45

Getting in a Flap

In ancient days two aviators procured to themselves wings.
Daedalus flew safely through the middle air and was duly
honoured on his landing. Icarus soared upwards to the sun till
the wax melted which bound his wings and his flight ended in
fiasco. The classical authorities tell us, of course, that he was
only 'doing a stunt'; but I prefer to think of him as the man
who brought to light a serious constructional defect in the
flying-machines of his day.

Arthur S. Eddington

Lots of things get around by flapping about: birds and butterflies
with their wings, whales and sharks with their tails, fish with their
fins. In all these situations there are three important factors at
work that determine the ease and efficiency of movement. First,
there is size – larger creatures are stronger and can have larger
wings and fins, which act on larger volumes of air or water. Next,
there is speed – the speed at which they can fly or swim tells us
how rapidly they are engaging with the medium in which they
are moving and the drag force that it exerts to slow them down.
Third, there is the rate at which they can flap their wings or fins.
Is there a common factor that would allow us to consider all the
different movements of birds and fish at one swoop?

As you have probably guessed, there is such a factor. When
scientists or mathematicians are faced with a diversity of examples

of a phenomenon, like flight or swimming, that differ in the detail but retain a basic similarity, they often try to classify the different examples by evaluating a quantity that is a pure number. By this I mean that it doesn't have any units, in the way that a mass or a speed (length per unit time) does. This ensures that it stays the same if the units used to measure the quantities are changed. So, whereas the numerical value of a distance travelled will change from 10,000 to $6\frac{1}{4}$ if you switch units from metres to miles, the ratio of two distances – like the distance travelled divided by the length of your stride – will not change if you measure the distance and your stride length *in the same units*, because it is just the number of strides that you need to take to cover the distance.

In our case there is one way to combine the three critical factors – the flapping rate per unit of time, f, the size of the flapping strokes, L, and the speed of travel, V – so as to get a quantity that is a pure number.* This combination is just fL/V and it is called the 'Strouhal number', after Vincenc Strouhal (1850–1922), a Czech physicist from Charles University in Prague.

In 2003 Graham Taylor, Robert Nudds and Adrian Thomas, at Oxford University, showed that if we evaluate the value of this Strouhal number St = fL/V for many varieties of swimming and flying animals at their cruising speeds (rather than in brief bursts when pursuing prey or escaping from attack), then they fall into a fairly narrow range of values that could be said to characterise the results of the very different evolutionary histories that led to these animals. They considered a very large number of different animals, but let's just pick on a few different ones to see something of this unity in the face of superficial diversity.

For a flying bird, f will be the frequency of wing flaps per second, L will be the overall span of the two flapping wings, and V will be the forward flying speed. A typical kestrel has an f of about 5.6

---

* The unit of frequency f is 1/time, of size L is length, and of speed V is length/time so the combination fL/V has no units: it is a pure dimensionless number.

flaps per second, a flap extent of about 0.34 metres and a forward speed of about 8 metres per second, giving St(kestrel) = (5.6 × 0.34)/8 = 0.24. A common bat has V = 6 metres per second, a wingspan of 0.26 metres, and a flapping rate of 8 times per second, so it has a Strouhal number of St(bat) = (8 × 0.26)/6 = 0.35. Doing the same calculation for forty-two different birds, bats and flying insects always gave a value of St in the range 0.2–0.4. They found just the same for the marine species they studied as well. More extensive studies were then carried out by Jim Rohr and Frank Fish (!) at San Diego and West Chester, Pennsylvania, to investigate this quantity for fish, sharks, dolphins and whales. Most (44 per cent) were found to lie in the range 0.23 to 0.28, but the overall range spanned 0.2 to 0.4, just like the range of values for the flying animals.

You can try this on humans too. A good male club standard swimmer will swim 100 metres in 60 seconds, so V = 5/3 metres per second, and uses about 54 complete stroke cycles of each arm (so the stroke frequency is 0.9 per second), with an arm-reach in the water of about 0.7 metres. This give St(human swimmer) = (0.9 × 2/3)/5/3 = 0.36, which places us rather closer to the birds and the fishes than we might have suspected. However, arguably the world's most impressive swimmer has been the Australian long-distance star Shelley Taylor-Smith, who has won the world marathon swimming championships seven times. She completed a 70 km swim in open sea water inside 20 hours with an average stroke rate of 88 strokes per minute. With an effective stroke reach of 1 metre that gives her the remarkable Strouhal number of 1.5, way up there with the mermaids.

46

Your Number's Up

Enter your postcode to view practical joke shops near you.

Practical Jokeshop UK

Life seems to be defined by numbers at every turn. We need to remember PIN numbers, account numbers, pass codes and countless reference numbers for every institution and government department under the Sun, and for several that never see it. Sometimes, I wonder whether we are going to run out of numbers. One of the most familiar numbers that labels us geographically (approximately) is the postcode. Mine is CB3 9LN and, together with my house number, it is sufficient to get all my mail delivered accurately, although we persist with adding road names and towns as back-up information or perhaps because it just sounds more human. My postcode follows an almost universal pattern in the United Kingdom: using four letters and two numbers. The positions of the letters and numbers don't really matter, although in practice they do because the letters also designate regional sorting and distribution centres (CB is Cambridge). But let's not worry about this detail – the postal service certainly wouldn't if it found itself running out of 6-symbol postcodes – and simply ask how many different postcodes of this form there be? You have 26 choices from A to Z for each of the four slots for letters and 10 choices from 0 to 9 for each of the numerals. If these are each chosen independently then the total number of different postcodes following the current pattern is equal to 26×26×10×10×26×26, which equals 45,697, 600 or nearly 46 million. Currently, the number of households in the United Kingdom is estimated to be about 26,222,000, or just over 26 million, and is projected to increase to about 28.5 million by the year 2020. So, even our relatively short postcodes have more than enough capacity to deal with the number of households and give each of them a unique identifier if required.

If we want to label individuals uniquely, then the postcode formula is not good enough. In 2006 the population of the United Kingdom was estimated at 60,587,000, about 60.5 million, and far greater than the number of postcodes. The closest thing that we have to an identity number is our National Insurance number, which is used by several agencies to identify us – the possibility of all these agencies coordinating their data by using this number is the step that alarms many civil liberty groups the most. A National Insurance number has a pattern of the form NA 123456 Z, which contains six numerals and 3 letters. As before, we can easily work out how many different National Insurance numbers this recipe permits. It's

$$26 \times 26 \times 10 \times 10 \times 10 \times 10 \times 10 \times 10 \times 26$$

This is a *big* number – 17,576,000,000 – seventeen billion, five hundred and seventy-six million, and vastly bigger than the population of the United Kingdom (and even than its projected value of 75 million by 2050). In fact, the population of the whole world is currently only about 6.65 billion and projected to reach 9 billion by the year 2050. So there are plenty of numbers – and letters – to go round.

47

Double Your Money

The value of your investments can go down as well as up.

UK consumer financial advice

Recently you will have discovered that the value of your invest-ments can plummet as well as go down. So, suppose you want to play safe and place cash in a straightforward savings account with a fixed, or slowly, changing rate of interest. How long will it take to double your money? Although nothing in this world is so certain as death and taxes (and the version of the latter that goes with the former), let's forget about them both and work out a handy rule of thumb for the time needed to double your money.

Start out by putting an amount A in a savings account with an annual fractional rate of interest r (so 5% interest corresponds to r = 0.05), then it will have grown to A × (1+r) after one year, to A × $(1+r)^2$ after two years, to A × $(1+r)^3$ after three years and so on. After, say, n years your savings will have become an amount equal to A × $(1+r)^n$. This will be equal to twice your original investment, that is 2A, when $(1+r)^n = 2$. If we take natural loga-rithms of this formula, and note that ln(2) = 0.69 approximately, and ln(1+r) is approximately equal to r when r is much less than 1 (which it always is – typically r is about 0.05 to 0.06 at present in the UK), then the number of years needed for your investment to double is given by the simple formula n = 0.69/r. Let's round

0.69 off to 0.7 and think of r as R per cent, so R = 100r, then we have the handy rule that[12]

$$n = 70/R$$

This shows, for example, that when the rate R is 7% we need about ten years to double our money, but if interest rates fall to 3.5% we will need twenty.

48

Some Reflections on Faces

In another moment Alice was through the glass, and had
jumped lightly down into the Looking-glass room.

Lewis Carroll

None of us has seen our own face – except in the mirror. Is the
image a true one? A simple experiment will tell you. After the bath-
room mirror has got steamed up, draw a circle around the image of
your face on the glass. Measure its diameter with the span of your
finger and thumb, and compare it with the actual size of your face.
You will always find that the image on the mirror is exactly one-half
of the size of your real face. No matter how far away from the mirror
you stand, the image on the glass will always be half size.

How strange this is. We have got so used to our appearance in
the mirror when shaving or combing our hair almost every day
of our lives that we have become immune to the big difference
between reality and perceived reality. There is nothing mysterious
about the optics of the situation. When we look at a plane mirror,
a 'virtual' image of our face is always formed at the same distance
'behind' the mirror as we are in front of it. Therefore, the mirror
is always located halfway between you and your virtual face. Light
does not pass through the mirror to create an image behind the
mirror, of course; it simply *appears* to be coming from this location. Walk towards a plane mirror and notice that your image
appears to approach you at twice the speed you are walking.

The next odd thing about your image in the mirror is that it has changed handedness. Hold your toothbrush in your right hand and it will appear in your left hand in the mirror image. There is a left–right reversal in the image, but the image is not inverted: there is no up–down reversal: if you are looking at your image in a hand-held mirror and you rotate the mirror clockwise by 90 degrees then your image is unchanged.

Hold up a transparent sheet with writing on it and something different happens. The writing is not reversed by the mirror if we hold the transparency facing us so that we can read it. If we held up a piece of paper facing the same way then we would not be able to read it in the mirror because it is, opaque. The mirror enables us to see the back of an object that is not trans-parent even though we are standing in front of it. But to see the front of it we would need to rotate it. If we rotate it around its vertical axis to face the mirror then we switch the right and left sides over. The left–right reversal of the mirror image is produced by this change of the object. If we rotate a page of a book that we are reading about its horizontal axis so that it faces the mirror then it appears upside down because it really has been inverted and not reversed in the left–right sense. When no mirror is present we don't get these effects because we can only see the front of the object we are looking at, and after it is rotated we only see the back. The reason the letters on the page of the book are reversed left–right is because we have turned the book about a vertical axis to make it face the mirror. The letters are not turned upside down, but we could make them appear upside down in the mirror if we turned the book about its horizontal axis, from bottom to top instead.

This is not the end of the story. There are some further inter-esting things that happen (as magicians know only too well) if you have two flat mirrors. Place them at right angles to make an L-shape and look towards the corner of the L. This is something that you can do with a dressing-table mirror with adjustable side mirrors.

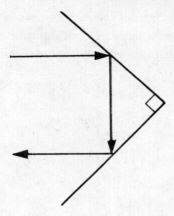

Look at yourself, or the pages of a book in this pair of right-angled mirrors and you will find that the image does *not* get swapped left–right. Your toothbrush appears to be in your right hand if it really is in your right hand. Indeed, to use such a mirror system for shaving or combing our hair is rather confusing because the brain automatically makes the left–right switch in practice. If you change the angle between the mirrors, gradually reducing it below 90 degrees, then something odd happens when you reach 60 degrees. The image looks just as it would when you look into a single flat mirror and is left–right reversed.

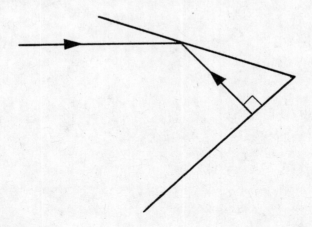

The 60-degree inclination of the mirrors ensures that a beam shone at one mirror will return along exactly the same path and create the same type of virtual image as you would see in a single plane mirror.

49

The Most Infamous Mathematician

He is the organiser of half that is evil and of nearly all that is undetected in [London]. He is a genius, a philosopher, an abstract thinker. He has a brain of the first order.

Sherlock Holmes in 'The Final Problem'

There was a time – and it may still be the time for some – when the most well-known mathematician among the general public was a fictional character. Professor James Moriarty was one of Arthur Conan Doyle's most memorable supporting characters for Sherlock Holmes. The 'Napoleon of crime' was a worthy adversary for Holmes and even required Mycroft Holmes's talents to be brought into play on occasion to thwart his grand designs. He appears in person only in two of the Holmes stories, 'The Final Problem' and 'The Valley of Fear', but is often lurking behind the scenes, as in the case of 'The Red-Headed League' where he plans an ingenious deception in order for his accomplices, led by John Clay, to tunnel into a bank vault from the basement of a neighbouring pawnshop.

We know a little of Moriarty's career from Holmes's descriptions of him. He tells us that

He is a man of good birth and excellent education, endowed by nature with a phenomenal mathematical faculty. At the age of twenty-one he

Professor James Moriarty

wrote a treatise upon the binomial theorem, which has had a European vogue. On the strength of it he won the mathematical chair at one of our smaller universities, and had, to all appearances, a most brilliant career before him.

But the man had hereditary tendencies of the most diabolical kind. A criminal strain ran in his blood, which, instead of being modified, was increased and rendered infinitely more dangerous by his extraordinary mental powers. Dark rumours gathered round him in the University town, and eventually he was compelled to resign his chair and come down to London . . .

Later, in *The Valley of Fear*, Holmes reveals a little more about Moriarty's academic career and his versatility. Whereas his early work was devoted to the problems of mathematical series, twenty-four years later we see him active in the advanced study of dynamical astronomy,

Is he not the celebrated author of *The Dynamics of an Asteroid*, a book which ascends to such rarefied heights of pure mathematics that it is said that there was no man in the scientific press capable of criticizing it?

Conan Doyle made careful use of real events and locations in setting his stories, and it is possible to make a good guess as to the real villain on whom Professor Moriarty was styled. The prime candidate is one Adam Worth (1844–1902), a German gentleman who spent his early life in America and specialised in audacious and ingenious crimes. In fact, a Scotland Yard detective of his day, Robert Anderson, did call him 'the Napoleon of the criminal world'. After starting out as a pickpocket and small-time thief, he graduated to organising robberies in New York. He was caught and imprisoned, but soon escaped and resumed business as usual, expanding its scope to include bank robberies and freeing the safe-breaker Charley Bullard from White Plains jail using a tunnel. Tellingly, for readers of 'The Red-Headed League', in November 1869, with Bullard's help he robbed the Boylston National Bank in Boston by tunnelling into the bank vault from a nearby shop. In order to escape the Pinkerton agents, Worth and Bullard fled to England and were soon carrying out robberies there and in Paris, where they moved in 1871. Worth bought several impressive properties in London and established a wide-ranging criminal network to ensure that he was always at arm's length from his robberies. His agents never even knew his name (he often used the assumed name of Henry Raymond), but it was impressed upon them that they should not use any violence in the perpetration of their crimes on his behalf. In the end, Worth was caught while visiting Bullard in prison and jailed for seven years in Leuven, Belgium, but was released in 1897 for good behaviour. He immediately stole jewellery to fund his return to normal life and, through the good offices of the Pinkerton detective agency in Chicago, arranged for the return of a painting, *The Duchess of Devonshire*, to the Agnew & Sons gallery in London in return for a 'reward' of $25,000. Worth then returned to London and lived there with his family until his death in 1902. His grave can be found in Highgate cemetery under the name of Henry J. Raymond.

In fact, Worth had stolen this painting of Georgiana Spencer (a great beauty and it appears, a relative through the Spencer family of Princess Diana) by Gainsborough* from Agnew's London gallery in 1876 and carried it around with him for many years, rather than sell it. It provides the key clue in establishing that Professor James Moriarty and Adam Worth were one and the same.

In *The Valley of Fear* Moriarty is interviewed by the police in his house. Hanging on the wall is a picture entitled '*La Jeune a l'agneau* – the young one has the lamb' – a pun on the 'Agnew' gallery that had lost the painting, although no one could ever prove that Worth had stolen it. But alas, as far as I can tell, Worth never wrote a treatise on the binomial theorem or a monograph on the dynamics of asteroids.

* The picture is now in the National Gallery of Art, Washington, D.C., and can be seen online at http://commons.wikimedia.org/wiki/Image:Thomas_Gainsboroguh_Georgiana_Duchess_o f_Devonshire_1783.jpg

50

Roller Coasters and Motorway Junctions

What goes up must come down.

Anon.

Have you ever been on one of those 'tear drop' roller coasters that take you up into a loop, over the top and back down? You might have thought that the curved path traces a circular arc, but that's almost never the case, because, if the riders are to reach the top with enough speed to avoid falling out of the cars at the top (or at least to avoid being supported only by their safety straps), then the maximum g-forces experienced by the riders when the ride returns to the bottom would become dangerously high.

Let's see what happens if the loop is circular and has a radius r and the fully loaded car has a mass m. The car will be gently started at a height h (which is bigger than r) above the ground and then descend steeply to the base of the loop. If we ignore any friction or air resistance effects on the motion of the car, then it will reach the bottom of the loop with a speed $V_b = \sqrt{2gh}$. It will then ascend to the top of the loop. If it arrives there with speed V_t, it will need an amount of energy equal to $2mgr + \frac{1}{2}mV_t^2$ in order to overcome the force of gravity and ascend a vertical height 2r to the top of the loop and arrive there with a speed V_t. Since the total energy of motion cannot be created or

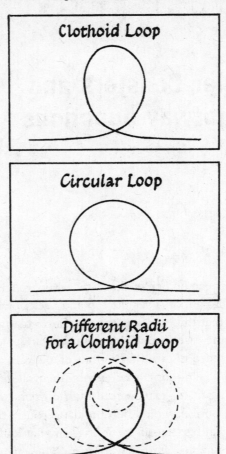

destroyed, we must have (the mass of the car m cancels out of every term)

$$gh = \tfrac{1}{2}\, V_b{}^2 = 2gr + \tfrac{1}{2}\, V_t{}^2$$

At the top of the circular loop the net force on the rider pushing upwards, and stopping him falling out of the car, is the force from the motion in a circle of radius r pushing upwards minus his weight pulling down; so, if the rider's mass is M, the

$$\text{Net upwards force at the top} = M\, V_t^2/r - Mg$$

This must be positive to stop him falling out, and so we must have $V_t^2 > gr$.

Looking back at the equations on pp.140–1, this tells us that we must have $h > 2.5r$. So if you just roll away from the start with the pull of gravity alone, you have got to start out at least 2.5 times higher than the top of the loop in order to get to the top with enough speed to avoid falling out of your seat. But this is a big problem. If you start that high up you will reach the bottom of the loop with a speed $V_b = \sqrt{(2gh)}$, which will be larger than $\sqrt{2g(2.5r)} = \sqrt{5gr}$. As you start to move in a circular arc at the bottom you will therefore feel a downward force equal to your weight *plus* the outward circular motion force, and this is equal to

$$\text{Net downwards force at the bottom} = Mg + MV_b^2/r > Mg + 5Mg$$

Therefore, the net downward force on the riders at the bottom will exceed six times their weight (an acceleration of 6-g). Most riders, unless they were off-duty astronauts or high-performance pilots wearing g-suits, would be rendered unconscious by this force. There would be no oxygen supply getting through to the brain at all. Typically, fairground rides with child riders aim to keep accelerations below 2-g, and those for adults experience at most 4-g.

Circular roller coaster rides seem to be a practical impossibility under this model, but if we look more closely at the two constraints – feel enough upward force at the top to avoid falling out but avoid experiencing lethal downward forces at the bottom – is there a way to change the roller coaster shape to meet both constraints?

When you move in a circle of radius R at speed V you feel an outward acceleration of V^2/R. The larger the radius of the circle and so the gentler the curve, the smaller the acceleration you will feel. On the roller coaster the V_t^2/r acceleration at the top is what is stopping us falling out, by overcoming our weight Mg acting downwards, so we want that to be big, which means r should be small at the top. On the other hand, when we are at the bottom the circular force is what is creating the extra 5-g of acceleration, and so we could reduce that by moving in a gentler circle with a larger radius. This can be achieved by making the roller coaster shape taller than it is wide, so it looks a bit like two parts of different circles, the one forming the top half with a smaller radius than the one forming the bottom half. The favourite curve that looks like this is called a 'clothoid' whose curvature decreases as you move along it in proportion to the distance moved. It was first introduced into roller coaster design in 1976 by the German engineer Werner Stengel for the 'Revolution' ride at Six Flags Magic Mountain in California.

Clothoids have another nice feature that has led to their incorporation into the design of complex motorway junction exits or railway lines. If a car is being driven along a curving motorway exit road, then as long as the driver keeps to a constant speed you can simply move the steering wheel with a constant rotation rate. If the bend were a different shape, then you would need to keep adjusting the rate of movement of the steering wheel or the speed of the car.

51

A Taylor-made Explosion

I do not know with what weapons World War III will be fought, but World War IV will be fought with sticks and stones.

Albert Einstein

The first atomic bomb exploded at the Trinity test in New Mexico, USA, 210 miles south of Los Alamos, on 16 July 1945. It was a watershed in human history. The creation of this device gave human beings the ability to destroy all human life and produce deadly long-term consequences. Subsequently, an arms race saw an escalation in the energetic yield of these bombs as the United States and the Soviet Union sought to demonstrate their ability to produce increasingly devastating explosions. Although only two of these devices were ever used in conflict,* the ecological and medical consequences of this era of tests in the atmosphere, on the ground, below ground and underwater, are still with us.

The explosions were much photographed at the time and produced a characteristic fireball and a canopy of debris that came

---

* The 'Little Boy' bomb dropped on Hiroshima was a fission device with 60 kg of uranium-235 and created a blast equivalent to 13 kilotons of TNT, killing approximately 80,000 people immediately; the 'Fat Man' bomb dropped on Nagasaki was a fission bomb with 6.4 kg of plutonium-239, equivalent to the blast from 21 kilotons of TNT. About 70,000 people were killed instantly.

to symbolise the consequences of nuclear war. The familiar mushroom cloud* forms for a reason. A huge volume of very hot gas with a low density is created at high pressure near ground level – atomic and nuclear bombs were generally detonated above ground to maximise the effects of the blast wave in all directions. Just like the bubbles rising in boiling water, the gas accelerates up into the denser air above, creating turbulent eddies curving downwards at their edges while additional debris and smoke streams up the centre in a rising column. The material at the core of the detonation is vaporised and heated to tens of millions of degrees, producing copious x-rays, which collide with and energise the atoms and molecules of the air above, creating a flash of white light whose duration depends on the magnitude of the initial explosion. As the front of the column rises, it spins like a tornado and draws in material from the ground to form the 'stem' of the growing mushroom shape; its density falls as it spreads, and eventually it finds itself with the same density as the air above it. At that moment, it stops rising and disperses sideways, so that all the material drawn up from the ground bounces backwards and descends to create a wide expanse of radioactive fall-out.

Ordinary explosions of TNT, or other non-nuclear weapons, have a rather different appearance because of the lower temperatures created at the start of a purely chemical explosion. This results in a turbulent mix of exploding gases rather than the organised stem and umbrella-like mushroom cloud.

One of the pioneers of the studies of the shape and character of large explosions was the remarkable Cambridge mathematician Geoffrey (G.I.) Taylor. Taylor wrote the classified report on the expected character of an atomic bomb explosion in June 1941. He became well known to a wider public after *Life* magazine in the

---

* The name 'mushroom cloud' became commonplace in the early 1950s, but the comparison between bomb debris patterns and 'mushrooms' dates back at least to newspaper headlines in 1937.

USA published a sequence of time-lapsed photographs of the 1945 Trinity test in New Mexico. The energy yield from this and other American atomic-bomb explosions was still top secret, but Taylor showed how a few lines of algebra enabled him (and hence anyone else with a smattering of simple mathematics) to work out the approximate energy of an explosion just by looking at the photographs.

Taylor was able to work out the expected distance to the edge of the explosion at any time after its detonation by noting that it can depend significantly on only two things: the energy of the explosion and the density of the surrounding air that it is ploughing through. There is only one way that such a dependence can look,[13] so approximately:

$$\text{Energy of bomb} = \frac{\text{air density} \times (\text{distance to explosion edge})^5}{(\text{time since detonation})^2}$$

The published photos showed the explosion at different times since detonation and those times were printed down the side of each photo with a distance scale running along the bottom of the photo to gauge the size. From the first frame of the photo Taylor noted that after 0.006 seconds the blast wave of the explosion had a radius of roughly 80 metres. We know that the density of air is 1.2 kilograms per cubic metre, and so the equation then tells us that the energy released was about 10^{14} Joules, which is equivalent to about 25,000 tons of TNT. For comparison, we know that the 2004 Indian earthquake released the energy equivalent to 475 million tons of TNT.

52

Walk Please, Don't Run!

You can spot the northern Europeans because they walk faster
than the leisurely art of the paseo strictly requires.

Benidorm travel guide

Walk along a busy city high street and most of the people around
you will be walking at about the same speed. A few people are in
a bit of a hurry and a few others move very slowly, perhaps because
of age, infirmity or totally impractical footwear. When you walk,
you keep a part of one foot in contact with the ground all the
time and you straighten your leg as you push off from the ground.
Indeed, the rules of race walking make these the defining features
of walking, which distinguish it from running: a failure to adhere
to them results in warnings and ultimately disqualification from
a race. As you walk, your hips will rise and fall as your centre
moves in a gentle circular arc each time you make a complete
stride. So if the length of your leg from the ground to your hips
is L, you will be creating an acceleration equal to v^2/L upwards
towards the centre of that circular arc of movement. This cannot
become greater than the acceleration due to gravity, g, that pushes
us down to the ground (or we would take off!), and so $g > v^2/L$
and we deduce that, roughly, the top speed for normal walking is
about $\sqrt{(gL)}$. Since $g = 10$ ms^{-2} and a typical leg length is 0.9 metre,
the top speed for ordinary walkers is about 3 metres per second
– a fairly good estimate – and the taller you are, the larger L will

be and the faster you will walk, although because of the square root there really isn't much difference between the walking speeds of people with the usual range of heights.

Another way to interpret this result is to look at people (or other two-legged creatures) trying to get from one place to another as quickly as possible and to ask at what speéd they stop walking and break into a run. The critical speed $\sqrt{(gL)}$ is the fastest rate of progress you can expect to make without breaking contact with the ground ('lifting' as the race walkers say). Once you start breaking contact you can go much faster, with a maximum speed of about $V = \sqrt{(2gnS)} \sim 9$ ms^{-1}, where $S \sim 0.3$ m is the difference in length between your straight leg and bent leg when you push off the ground and $n \sim 10$ is the number of strides you use to accelerate up to full speed.

Race walkers walk much faster than 3 metres per second. The world record for walking 1,500 metres, set by American Tim Lewis in 1988, is 5 minutes 13.53 seconds, an average speed of 4.78 ms^{-1}. This event is rarely walked, and so it is interesting to look at the highly competitive world record for 20 kilometres, the shorter of the two championship events. This was reduced to 1 hr 17 mins and 16 secs by the Russian walker, Vladimir Kanaykin, on 29 September 2007, an average speed of 4.3 ms^{-1} over more than 12.5 miles! These speeds manage to exceed our estimate of $\sqrt{(gL)}$ because race walkers use a much more efficient style of walking than we do when we stroll down the road. They do not rock their centres up and down and are able to roll their hips in a very flexible fashion to produce a longer stride length and higher frequency of striding. This highly efficient movement, coupled with very high levels of fitness, enables them to sustain impressive speeds over long periods of time. The world record holder over 50 km, more than 31 miles, averages more than 3.8 ms^{-1} and will cover the marathon distance (42.2 km) in 3 hours 6 minutes en route.

53

Mind-reading Tricks

Every positive integer is one of Ramanujan's personal friends.

John E. Littlewood

Think of a number between 1 and 9. Multiply it by 9 and add the digits of this new number together. Subtract 4 from your answer and you will be left with a single-digit number. Next, convert this number to a letter: if your number is 1 it becomes A, 2 becomes B, 3 becomes C, 4 becomes D, 5 becomes E, 6 becomes F and so on. Now think of a type of animal that begins with your chosen letter and imagine that animal as strongly as you can. Hold it vividly in the forefront of your mind. If you look at the note 14 at the back of the book, you will see that I have read your mind and discovered the animal that you were imagining.

This is a very simple trick, and you ought to be able to work out how I was able to guess the animal of your choice with such a high likelihood of success. There is a little mathematics involved, in that some simple properties of numbers are exploited, but there is also a psychological – and even zoological – ingredient as well.

There is another trick of this general sort that involves only the properties of numbers. It uses the number 1089, which you may well already have listed among your favourites. It was the year in which there was an earthquake in England; it is also a perfect square (33×33); but its most striking property is the following.

Pick any three-digit number in which the digits are all different (like 153). Make a second number by reversing the order of the three digits (so 351). Now take the smaller of the two numbers away from the larger (so $351 - 153 = 198$; if your number has only two digits, like 23, then put a 0 in front, so 023). Now add this to the number you get by writing it backwards (so $198 + 891 = 1089$). Whatever number you chose at the outset, you will end up with 1089 after this sequence of operations![15]

54

The Planet of the Deceivers

You can fool some of the people some of the time, and some of the people all the time, but you cannot fool all the people all of the time.

Abraham Lincoln

One of the human intuitions that has been honed by countless generations of social interaction is trust. It is founded upon an ability to assess how likely it is that someone is telling the truth. One of the sharp distinctions between different environments is whether we assume people are honest until we have reason to think otherwise, or whether we assume them to be dishonest until we have reason to think otherwise. One encounters this distinction in the bureaucracies of different countries. In Britain officialdom is based upon the premise that people are assumed to be honest, but I have noticed that in some other countries the opposite is the default assumption and rules and regulations are created under a presumption of dishonesty. When you make an insurance claim, you will discover which option your company takes in its dealings with its customers.

Imagine that our quest to explore the Universe has made contact with a civilisation on the strange world of Janus. The monitoring of their political and commercial activity over a long period has shown us that on the average their citizens tell the truth ¼ of the time and lie ¾ of the time. Despite this worrying appraisal, we decide to go ahead with a visit and are welcomed by the leader of their majority party who makes a grand statement about his benevolent intentions. This is followed by the leader of the opposition party getting up and saying that the Leader's statement was a true one. What is the likelihood that the Leader's statement was indeed a true one?

We need to know the probability that the Leader's statement was true given that the opposition head said it was. This is equal to the probability that the Leader's statement was true and the opposition head's claim was true divided by the probability that the opposition head's statement was true. Well, the first of these – the probability that they were both telling the truth is just $1/4 \times 1/4 = 1/16$. The probability that the opponent spoke the truth is the sum of two probabilities: the first is the probability that the Leader told the truth and his opponent did too, which is $1/4 \times 1/4 = 1/16$, and the probability that the Leader lied and his opponent lied too, which is $3/4 \times 3/4 = 9/16$. So, the probability[16] that the Leader's statement was really true was just $1/16 \div 10/16$, that is $1/10$.

55

How to Win the Lottery

A lottery is a taxation – upon all the fools in creation. And heaven be praised, it is easily raised, credulity's always in fashion.

Henry Fielding

The UK National Lottery has a simple structure. You pay £1 to select six different numbers from the list 1,2,3, . . . , 48,49. You win a prize if at least three of the numbers on your ticket match those on six different balls selected by a machine that is designed to make random choices from the 49 numbered balls. Once drawn, the balls are not returned to the machine. The more numbers you match, the bigger the prize you win. Match all six and you will share the jackpot with any others who also share the same six matching numbers. In addition to the six drawn balls, an extra one is drawn and called the 'Bonus Ball'. This affects only those players who have matched five of the six numbers already drawn. If they also match the Bonus Ball then they get a larger prize than those who matched only the other five numbers.

What are your chances of picking six numbers from the 49 possibilities correctly, assuming that the machine* picks winning

* Strictly speaking there are 12 machines (which each have names) and 8 sets of balls that can be used for the public draw of the winning numbers on television. The machine and the set of balls to be used at any draw are chosen at random

numbers at random? The drawing of each ball is an independent event that has no effect on the next drawing, aside from reducing the number of balls to be chosen from. The chance of getting the first of the 6 winning numbers from the 49 is therefore just the fraction 6/49. The chance of picking the next of the remaining 5 from the 48 balls that remain is 5/48. The chance of picking the next of the remaining 4 from the 47 balls that remain is 4/47. And so on, the remaining three probabilities being 3/46, 2/45 and 1/44. So the probability that you pick them all independently and share the jackpot is

$$6/49 \times 5/48 \times 4/47 \times 3/46 \times 2/45 \times 1/44 = 720/10068347520$$

If you divide this out you get the odds as 1 in 13,983,816 – that's about one chance in 13.9 million. If you want to match 5 numbers plus the Bonus Ball, then the odds are 6 times smaller, and your chance of sharing the prize is 1 in 13,983,816/6 or 1 in 2,330,636.

Let's take the collection of all the possible draws – all 13,983,816 of them – and ask how many of them will result in 5, or 4, or 3, or 2, or 1, or zero numbers being chosen correctly.[17] There are just 258 of them that get 5 numbers correct, but 6 of them win the Bonus Ball prize, so that leaves 252; 13,545 of them get 4 balls correct, 246,820 of them that get 3 balls correct, 1,851,150 of them that get 2 balls correct, 5,775,588 of them get just 1 ball correct, and 6,096,454 of them get none of them correct. So to get the odds for you to get, say, 5 numbers correct you just divide the number of ways it can happen by the total number of possible

---

from these candidates. This point is usually missed by those who carry out statistical analyses of the results of the draw since the Lottery began. Since the most likely source of a non-random element favouring a particular group of numbers over others would be associated with features of a particular machine or ball, it is important to do statistical studies for each machine and set of balls separately. Such biases would be evened out by averaging the results over all the sets of balls and machines.

combinations, i.e. 252/13,983,816, which means odds of 1 in 55,491 if you buy one lottery ticket. For matching 4 balls the odds are 1 in 1,032; for matching 3 balls they are 1 in 57. The number of the 13,983,816 outcomes that win a prize is 1 + 258 + 13,545 + 246,820 = 260,624 and so the odds of winning any prize when you buy a single ticket are 1 in 13,983,816/260,624, that is about 1 in 54. Buy a ticket a week with an extra one on your birthday and at Christmas and you have an evens chance of winning something.

This arithmetic is not very encouraging. Statistician John Haigh points out that the average person is more likely to drop dead within one hour of purchasing a ticket than to win the jackpot. Although it is true that if you don't buy a ticket you will certainly not win, what if you buy lots of tickets?

The only way to be sure of winning a lottery is to buy *all* the tickets. There have been several attempts to use such a strategy in different lotteries around the world. If no jackpot is won in the draw, then usually the unwon prize is rolled over to the following week to create a super-jackpot. In such situations it might be attractive to try to buy almost all the tickets. This is quite legal! The Virginia State Lottery in the USA is like the UK Lottery except the six winning numbers are chosen from only 44 balls, so there are 7,059,052 possible outcomes. When the jackpot had rolled over to $27 million, Australian gambler Peter Mandral set in operation a well-oiled ticket buying and printing operation that managed to buy 90% of the tickets (a failure by some of his team was responsible for the worrying gap of 10%). He won the rollover jackpot and went home with a healthy profit on his $10 million outlay on tickets and payments to his ticket-buying 'workers'.

56

A Truly Weird Football Match

Own goal: Own goals tend, like deflections, to be described with sympathy for those who fall victim to them. Often therefore preceded by the adjectives *freak* or *bizarre* even when 'incompetent' or 'stupid' might come more readily to mind.

John Leigh and David Woodhouse, *The Football Lexicon*

What is the most bizarre football match ever played? In that competition I think there is only one winner. It has to be the infamous encounter between Grenada and Barbados in the 1994 Shell Caribbean Cup. This tournament had a group stage before the final knockout matches. In the last of the group stage games Barbados needed to beat Grenada by at least two clear goals in order to qualify for the next stage. If they failed to do that, Grenada would qualify instead. This sounds very straightforward. What could possibly go wrong?

Alas, the law of unforeseen consequences struck with a vengeance. The tournament organisers had introduced a new rule in order to give a fairer goal difference advantage to teams that won in extra time by scoring a 'golden goal'. Since the golden goal ended the match, you could never win by more than one goal in such a circumstance, which seems unfair. The organisers therefore decided that a golden goal would count as two goals. But look what happened.

Barbados soon took a 2–0 lead and looked to be coasting through to the next phase. Just seven minutes from full time Grenada pulled a goal back to make it 2–1. Barbados could still qualify by scoring a third goal, but that wasn't so easy with only a few minutes left. Better to attack their own goal and score an equaliser for Grenada because they then had the chance to win by a golden goal in extra time, which would count as two goals and so Barbados would qualify at Grenada's expense. Barbados obliged by putting the ball into their own net to make it 2–2 with three minutes left. Grenada realised that if they could score another goal (at either end!) they would go through, so they attacked their own goal to get that losing goal that would send them through on goal difference. But Barbados resolutely defended the Grenada goal to stop them scoring and sent the match into extra time. In extra time the Barbadians took their opponents by surprise by attacking the Grenada goal and scored the winning golden goal in the first five minutes. If you don't believe me, watch it on YouTube!*

* http://www.youtube.com/watch?v=ThpYsN-4p7w

57

An Arch Problem

Genius is four parts perspiration and one part having a focused strategic overview.

Armando Iannucci

An old arch of stones can seem a very puzzling creation. Each stone looks as if it has been put in place individually, but the whole structure looks as if it cannot be supported until the last capstone is put in place: you can't have an 'almost' arch. So, how could it have been made?

The problem is an interesting one because it is reminiscent of a curious argument that is much in evidence in the United States under the name of 'Intelligent Design'. Roughly speaking, its advocates pick on some complicated things that exist in the natural world and argue that they must have been 'designed' in that form rather than have evolved by a step-by-step process from simpler forms because there is no previous step from which they could have developed. This is a little subjective, of course – we may not be very imaginative in seeing what the previous step was – but at root the problem is just like our arch, which is a complicated construct that doesn't seem to be one step away from a slightly simpler version of an arch with one stone missing.

Our unimaginative thinking in the case of the arch is that we have got trapped into thinking that all structures are built up by adding bits to them. But some structures can be built by subtraction. Suppose we started with a heap of stones and gradually shuffled them and removed stones from the centre of the pile until we left an arch behind. Seen in this way we can understand what the 'almost' arch looks like. It has part of the central hole filled in. Real sea arches are made by the gradual erosion of the hole until only the outer arch remains. Likewise, not all complexity in Nature is made by addition.

58

Counting in Eights

> The Eightfold Path: Right view, right intention, right speech,
> right action, right livelihood, right effort, right mindfulness
> and right concentration.

> The Noble Eightfold Way

We count in 'tens'. Ten ones make ten, ten tens make a hundred,
ten hundreds make a thousand and so on forever. This is why our
counting system is called a 'decimal' system. There is no limit to
its scope if you have enough labels to name the results. We have
words like million, billion and trillion for some of the large
numbers, but not for every one that you might need to write
down. Instead we have a handy notation that writes 10^n to denote
the number which is 1 followed by n noughts, so a thousand is
10^3.

The origin of all these tens in the counting system is not hard
to find. It is at our fingertips. Most ancient human cultures used
their fingers in some way for counting. As a result you find counting
systems based on groups of five (fingers of one hand), ten (fingers
of both hands), groups of twenty (fingers plus toes), or mixtures
of all or some of these systems. Our own counting system betrays
a complicated history in which different counting systems merged
to form new ones by the presence of old words that reflect the
previous base. Thus we have a word like 'dozen', for 12, or 'score'
(derived from the old Saxon word *sceran*, meaning to shear or to

cut) for 20, with its interesting triple meaning of 20, to make a mark or to keep count. All three meanings reflect the time when tallies were kept on pieces of wood by marking (scoring) them in groups of 20.

Despite the ubiquity of the base 10 counting system in early culture, there is one unusual case where a Central American Indian society used a base 8 counting system. Can you think why this might be? I used to ask mathematicians if they could think of a good reason, and they usually responded by saying that 8 was a good number, to use because it has lots of factors, it divides exactly by 2 and 4, so you can divide portions into quarters without creating a new type of quantity that we call a fraction. The only time I got the right answer though was when I asked a large group of 8-year-old children and one girl immediately produced the answer: they were counting the gaps between their fingers. If you hold things between your fingers, strings or pieces of material, this is a natural way to count. The base eighters were finger counters too.

59

Getting a Mandate

Democracy used to be a good thing, but now it has gotten
into the wrong hands.

Jesse Helms

Politicians have a habit of presuming that they have a much greater
mandate than they really do. If you have a roster of policies on
offer to the electorate, the fact that you gained the most voters
overall does not mean that a majority of voters favour each one
of your policies over your opponent's alternative. And if you win
the election by a narrow margin, what sort of mandate do you
have?

For simplicity, assume there are just two candidates (or parties)
in the election. Suppose the winner gets W votes and the loser
gets L votes, so the total number of valid votes cast was W+L.
In any number of events of this size, the random statistical 'error'
that you expect to occur is given by the square root of W+L, so
if W+L = 100 there will be a statistical uncertainty of 10 in either
direction. In order for the winner of the election to be confident
that they haven't won because of a significant random variation
in the whole voting process – counting, casting and sorting votes
– we need to have the winning margin greater than the random
variation:

$$W-L > \sqrt{(W+L)}$$

If 100 votes were cast, then the winning margin needs to exceed 10 votes in order to be convincing. As an example, in the 2000 US Presidential election* Bush received 271 electoral college votes and Gore received 266. The difference was just 5, far less than the square root of 271 + 266, which is about 23.

More amusingly, it is told that Enrico Fermi, the great Italian high-energy physicist who was a key player in the creation of the first atomic bomb – and a very competitive tennis player – once responded to being beaten at tennis by 6 games to 4 by remarking that the defeat was not statistically significant because the margin was less than the square root of the number of games they played!

Let's suppose you have won the election and have a margin of victory large enough to quell concerns about purely random errors being responsible, how large a majority in your favour do you think you need in order to claim that you have a 'mandate' from the electorate for your policies? One interesting suggestion is to require that the fraction of all the votes that the winner receives, $W/(W+L)$, exceeds the ratio of the loser's votes to the winner's votes, L/W. This 'golden' mandate condition is therefore that

$$W/(W+L) > L/W$$

This requires that $W/L > (1+\sqrt{5})/2 = 1.61$, which is the famous 'golden ratio'. This means that you would require a fraction $W/(W+L)$ greater than 8/13 or 61.5% of all the votes cast for the two parties. In the last general election in the UK, Labour won 412 seats and the Conservatives 166, so Labour had 71.2% of these

---

* Some aspects of this election remain deeply suspicious from a purely statistical point of view. In the crucial Florida vote the result of the recount was very mysterious. Just re-examining ballot papers produced a gain of 2,200 votes for Gore and 700 for Bush. Since one would expect there to be an equal chance of an ambiguous ballot being cast for either candidate, this huge asymmetry in the destination of the new ballots accepted in the recount suggests that something else of a non-random nature was going on in either the first count or the recount.

578 seats, enough for a 'golden' mandate. By contrast, in the 2004 US Election, Bush won 286 electoral votes, Kerry won 251, and so Bush received only 53.3% of the total, less than required for a 'golden' mandate.

60

The Two-headed League

But many that are first shall be last; and the last shall be first.

Gospel of St. Matthew

In 1981 the Football Association in England made a radical change to the way its leagues operated in an attempt to reward more attacking play. They proposed that 3 points be awarded for a win rather than the 2 points that had traditionally been the victor's reward. A draw still received just 1 point. Soon other countries followed suit, and this is now the universal system of point scoring in football league competitions all over the world. It is interesting to look at the effect this has had on the degree of success that a dull, non-winning team can have. In the era of 2 points for a win it was easily possible to win the league with 60 points from 42 games, and so a team that gained 42 points from drawing all its games could finish in the top half of the league – indeed, Chelsea won the old First Division Championship in 1955 with the lowest ever points total of 52. Today, with 3 points for a win, the champion side needs over 90 points from its 38 games and an all-drawing side will find its 42 points will put it three or four from the bottom, fighting against relegation.

With these changes in mind, let's imagine a league where the football authorities decide to change the scoring system just after the final whistles blow on the last day of the season. Throughout the season they have been playing 2 points for a win and 1 point for a draw. There are 13 teams in the league and they play each other once, so every team plays 12 games. The All Stars win 5 of their games and lose 7. Remarkably, every other game played in the league is drawn. The All Stars therefore score a total of 10 points. All the other teams score 11 points from their 11 drawn games, and 7 of them score another 2 points when they beat the All Stars, while 5 of them lose to the All Stars and score no more points. So 7 of the other teams end up with 13 points and 5 of them end up with 12 points. All of them have scored more than the All Stars, who therefore find themselves languishing at the bottom of the league table.

Just as the despondent All Stars have got back to the dressing room after their final game and realise they are bottom of the league, facing certain relegation and probable financial ruin, the news filters through that the league authorities have voted to introduce a new points scoring system and apply it retrospectively to all the matches played in the league that season. In order to reward attacking play they will award 3 points for a win and 1 for a draw. The All Stars quickly do a little recalculating. They now get 15 points from their 5 wins. The other teams get 11 points from their 11 drawn games still. But now the 7 that beat the All Stars only get another 3 points each, while the 5 that lost to them get nothing. Either way, all the other teams score only 11 points or 14 points and the All Stars are now the champions!

61

Creating Something out of Nothing

Mistakes are good. The more mistakes, the better. People who make mistakes get promoted. They can be trusted. Why? They're not dangerous. They can't be too serious. People who don't make mistakes eventually fall off cliffs, a bad thing because everyone in free fall is considered a liability. They might land on you.

James Church, *A Corpse in the Koryo*

If you are one of those people who have to give lectures or 'presentations' using a computer package like PowerPoint, then you have probably also discovered one of its weaknesses – especially if you are an educator. When your presentation ends it is typically time for questions from the audience about what you have said. One of the facts of life about such questions is that they are very often most easily and effectively answered by drawing or writing something. If you are at a blackboard or have an overhead projector with acetate sheets and a pen in front of you, then any picture you need is easily drawn. But armed just with your standard laptop computer you are a bit stuck. You can't easily 'draw' anything over your presentation unless you have a 'tablet PC'. Which all goes to show how much we rely on pictures to explain what we mean. They are more direct than words. They are analog, not digital.

Some mathematicians are suspicious of drawings. They like proofs that make no reference to a picture you might have drawn in a way that biases what you think might be true. But most mathematicians are quite the opposite. They like pictures and see them as a vital guide to seeing what might be true and how to go about showing it. Since that opinion is in the majority, let's show something that will make the minority happy. Suppose you have 8 × 8 square metres of expensive flooring that is made up of four pieces – two triangles and two quadrilaterals – as shown in the floor plan here.

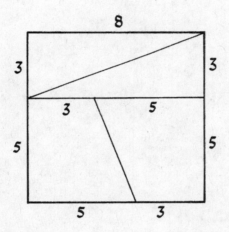

It is easy to see that the total area of the square is 8 × 8 = 64 square metres. Now let's take our four pieces of flooring with the given dimensions and lay them down in a different way. This time to create a rectangle, like this:

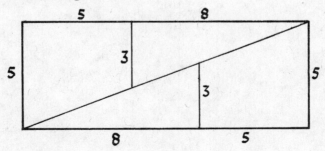

Something strange has happened though. What is the area of the new rectangular carpet? It is $13 \times 5 = 65$ square metres.[18] We have created one square metre of carpet out of nothing! What has happened? Cut out the pieces and try it for yourself.

62

How to Rig An Election

I will serve as a consultant for your group for your next election. Tell me who you want to win. After talking to the members I will design a 'democratic procedure' which ensures the election of your candidate.

Donald Saari[19]

As we have already seen in Chapter 14, elections can be tricky things. There are many ways to count votes, and if you do it unwisely you can find that candidate A beats B who beats C who loses to A! This is undesirable. Sometimes we find ourselves voting several times on a collection of candidates with the weakest being dropped at each stage, so that the votes for that candidate can be transferred to other candidates in the next round of voting.

Even if you are not spending your days casting formal votes for candidates, you will be surprised how often you are engaged in voting. What film shall we go to see? What TV channel shall we watch? Where shall we go on holiday? What's the best make of fridge to buy? If you enter into discussion with others about questions that different possible answers then you are really engaged in casting a 'vote' – your preference – and the successful 'candidate' is the choice that is finally made. These decisions are not reached by all parties casting a vote. The process is usually much more haphazard. Someone suggests one film. Then someone suggests another because it is newer. Then someone says the newer one is too violent and we should pick a third one. Someone turns out to have seen that one already so it's back to the first choice. Someone realises it's no good for children so they suggest another. People are wearying by now and agree on that proposal. What is happening here is interesting. One possibility at a time is being considered against another one, and this process is repeated like rounds in a tournament. You never consider all the attributes of all the possible films together and vote. The outcome of the deliberations therefore depend very strongly on the order in which you consider one film versus another. Change the order in which you consider the films and the attributes you use for comparing them and you can end up with a very different winner.

It's just the same with elections. Suppose you have 24 people who have to choose a leader from 8 possible candidates (A,B,C,D,E,F,G and H). The 'voters' divide up into three groups who decide on the following orderings of their preferences among the candidates:

> 1st group: A B C D E F G H
> 2nd group: B C D E F G H A
> 3rd group: C D E F G H A B

At first glance it looks as if C is the preferred candidate overall, taking 1st, 2nd and 3rd places in the three ranking lists. But H's mother is very keen for H to get the leader's job and comes to ask if we can make sure that H wins the vote. It looks hopeless as H is last, second last and third last on the preference lists. There must be no chance of H becoming leader. We make it clear to H's mother that everything must follow the rules and no dishonesty is allowed. So, the challenge is to find a voting system that makes H the winner.

All we need to do in order to oblige is set up a tournament-style election and pick the winner of each 2-person contest using the preferences of the three groups listed above. First, pit G against

F, and we see F wins 3–0. F then plays E, and loses 3–0. E then plays D and loses 3–0. D then plays C and loses 3–0. C then plays B and loses 2–1. B then plays A and loses 2–1. That leaves A to play H in the final match up. H beats A 2–1. So H is the winning choice in this 'tournament' to decide on the new leader.

What was the trick? Simply make sure the stronger candidates eliminate each other one by one in the early stages and introduce your 'protected' candidate only at the last moment, ensuring that they are compared only with other candidates they can beat. So, with the right seeding, a British tennis player can win Wimbledon after all.

63

The Swing of the Pendulum

England swings like a pendulum do.

Roger Miller

The story is told that in the sixteenth century the great Italian scientist Galileo Galilei used to amuse himself by watching the swinging of the great bronze chandelier that hung from the ceiling of the cathedral in Pisa. It may have been set in motion to spread the aroma of incense, or it may have been disturbed by the need to lower it to replenish the candles. What he saw fascinated him. The rope suspending the chandelier from the ceiling was very long and so the chandelier swung backwards and forwards like a pendulum, very slowly: slow enough to time how long it would take to make a complete journey out and back to its starting point. Galileo observed what happened on many occasions. On each occasion the chandelier swung differently; sometimes making only very small swings; sometimes somewhat larger ones – but he had noticed something very important. The period of time taken by the swinging chandelier to complete a single out and back swing was the same regardless of how far it swung. Given a large push, it went further than if it were given a small one. But if it went further, it went faster and took the same time to get back to where it started from as if it were pushed very gently.

This discovery* has far-reaching consequences. If you have a grandfather clock then you will have to wind it up about once a week. Galileo's discovery means that if the pendulum has stopped, it doesn't matter how your push starts it swinging again. As long as the push is not too big, it will take the same time to swing back and forth and the resulting 'tick' will have the same duration. Were it not so, then pendulum clocks would be very tedious objects. You would have to set the amplitude of the swinging pendulum exactly right in order that the clock keeps the same time as it did before it stopped. Indeed, Galileo's acute observation is what led to the idea of the pendulum clock. The first working version was made by the Dutch physicist Christiaan Huygens in the 1650s.

Finally, there is a nice test of whether a physicist's logic is stronger than his survival instinct that exploits the swing of a pendulum. Imagine that we have a huge, heavy pendulum, like the cathedral chandelier that Galileo watched. Stand to one side of it and draw the pendulum weight towards you until it just touches the tip of your physicist's nose. Now just let the pendulum weight go. Don't give it any push. It will swing away from your nose and then return back towards the tip of your nose. Will you flinch? Should you flinch? The answers are, well, 'Yes and no'.†

---

* Galileo thought it was true for all swings of the chandelier, no matter how far they went. In fact, that is not true. It is true to very high accuracy for oscillations of 'small' amplitude. This type of oscillation is known to scientists as 'simple harmonic motion'. It describes the behaviour of almost every stable system in Nature after it is slightly perturbed from its equilibrium state.

† The pendulum cannot swing back to a greater height than where it began from (unless someone hits it to give it more energy). In practice, the pendulum always loses a little of its energy overcoming air resistance and overcoming friction at the point of its support, and so it will never quite return to the same height that it began from. The physicist is actually quite safe – but will always flinch none the less.

64

A Bike with Square Wheels

The bicycle is just as good company as most husbands and, when it gets old and shabby, a woman can dispose of it and get a new one without shocking the entire community.

Ann Strong

If your bike's like my bike it's got circular wheels. You might just have one of them, but you've probably got two of them, but either way they are circular. However, you will probably be surprised to learn that it didn't have to be like this. You can have a perfectly steady ride on a bike with square wheels, as long as you ride on the right type of surface.

The important feature of a rolling wheel for the cyclist is that you don't keep jolting up and down as the bicycle moves forward. With circular wheels on a flat surface this is the case: the centre of the cyclist's body moves forward in a straight line when the bicycle moves forward in a straight line without any slipping. Use square wheels on a flat surface and you are going to have a very uncomfortable up-and-down ride. But could there be a differently shaped road surface that results in a smooth ride when you are using square wheels? All we need to check is whether there is a shape that leads to a straight-line motion of the cyclist when he has square wheels.

The answer is pretty surprising. The shape of the road surface that gives a steady ride on square wheels is created when you hang two ends of a chain from two points at the same height above ground. This is called the catenary and we have met it before in that connection in Chapter 11. If we turn it upside down we obtain the shape that is used for many of the great arches of the world. But if you take a catenary arch and repeat it over and over again along a line you get a sequence of undulations of the same height. This is the shape of the ground surface that gives us the smooth ride on square wheels. We just need the bottom corners of the square to fit into the sides of the successive 'valleys' on the surface. The key feature of the catenary is that when one is placed next to another, the angle between the two sides of the adjacent arches as they come into the lowest point of each valley is a right angle, 90 degrees, and that is the angle at each corner of the square. So, the right-angled wheel just keeps rolling along.*

---

* The square is not the only possible wheel shape that can give a smooth ride. Any polygonally-shaped wheel will work for a different catenary-shaped road. As the number of sides in the polygon gets very large, it begins to look more and more like a circle and the line of catenaries gets flatter and flatter and looks increasingly like a perfectly flat road. The case of a 3-sided polygon, a triangular wheel, is problematic because it runs into the side of the upcoming catenary before it can roll into the corner and hits the trailing side as it rolls out. You need to lay the road surface bit by bit to avoid these collisions happening. For a rolling polygonal wheel with N equal sides (so N = 4 is the case of our squared wheel), the equation of the catenary-shaped road that gives a smooth, straight-line rider for the cyclist is $y = -B \cosh(x/B)$, where $B = C \cot(\pi/N)$ with C constant.

65

How Many Guards Does an Art Gallery Need?

Who will guard the guards?

Juvenal

Imagine you are head of security at a large art gallery. You have many valuable paintings covering all of the gallery walls. They are also quite low on the walls, so that they can be viewed at eye level and so are vulnerable to theft or vandalism. The gallery is a collection of different rooms with different shapes and sizes. How are you going to make sure that each one of the pictures can be kept under surveillance all of the time. The solution is simple if you have unlimited money: just have one attendant standing on guard next to every picture. But art galleries are rarely awash with money, and wealthy donors don't tend to earmark their gifts for the provision of guards and their chairs. So, in practice, you have a problem, a mathematical problem: what is the smallest number of attendants that you need to hire, and how should you position them so that all the walls of the gallery are visible at eye level?

We need to know the minimum number of guards needed to watch N walls. We assume that the walls are straight and a guard at a corner where two walls meet will be assumed to be able to see everything on those two walls, and we will assume that a guard's view is never obstructed. A square gallery can obviously be watched by just one guard. In fact, if the gallery is shaped like any polygon in which all the sides bulge outwards (a 'convex' polygon) then one guard will always suffice.

Things get more interesting when the walls don't all bulge outwards. Here is a gallery like that, with 8 walls, which can also be guarded by just one attendant located at the corner O.

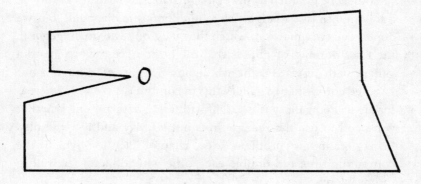

So, this is a rather economical gallery to run.

Here is another 'kinkier' 12-walled gallery that is not so efficient. It needs 4 guards to keep an eye on all the walls.

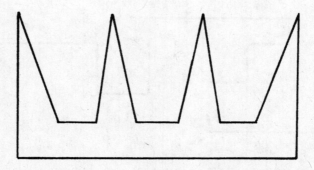

To solve the problem in general we just look at how we can divide the gallery up into triangles that don't overlap. This can always be done. If the polygon has S vertices, then there will be S-2 triangles. Since a triangle is one of those convex shapes (the three-sided one) that needs only a single guard we know that if the gallery can be completely covered by, say, T non-overlapping triangles then it can always be guarded by T attendants. It might, of course, be guarded by fewer. For instance, we can always divide a square into two triangles by joining opposite diagonals, but we don't need two guards to watch the walls, one will do. In general, the total number of guards that might be necessary to guard a gallery with W walls is the whole number part of W/3. For our 12-sided comb-shaped gallery this maximum is 12/3 = 4, whereas for an 8-sided gallery it is 2. Unfortunately, determining whether you need to use the maximum is not so easy and is a so-called 'hard' computer problem (See chapter 27) for which the computing time can double each time you add another wall to the problem.

Most of the galleries you visit today will not have kinked, jagged wall plans like these examples. They will have walls that are all at right angles like this:

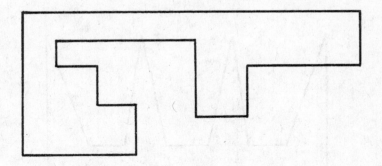

If there are many corners in a right-angled polygonal gallery like this, then the number of attendants located at the corners that might be necessary and is always sufficient to guard the gallery is the whole number part of ¼ × (Number of Corners). For the 14-cornered gallery shown here this number is 3. This means that it is much more economical on wages to have a gallery design like this, especially when it is large. If you have 150 walls, then the non-right-angled design could need 50 guards, while the right-angled set-up will need at most 37.

Another traditional type of right-angled gallery will be divided into rooms, like this 10-roomed example:

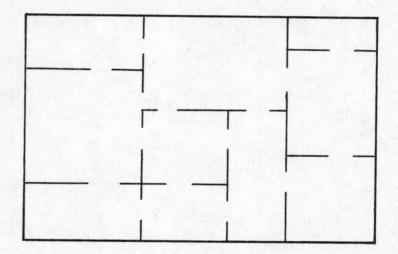

In these cases you can always divide the galley up into a collection of non-overlapping rectangles. This is a useful design because if you place an attendent at the opening connecting two different rooms, then he can guard both of them at the same time. But no guard can guard 3 or more rooms at once. So, now the number of guards that is sufficient and occasionally necessary to keep a complete watch on the gallery is the whole number part of ½ × (Number of Rooms), or 5 for the gallery drawn here. This is a very economical use of human resources.

We have been speaking of people watching the walls, but the things we have said also apply to CCTV cameras or to lights for illuminating the gallery and its rooms. Next time you are planning to steal the *Mona Lisa* you will have a head start.

66

. . . and What About a Prison?

All my contacts with criminals show that what they're doing is just a slightly more extreme version of what everybody is doing.

David Carter

Art galleries are not the only buildings that have to be guarded. Prisons and castles do too. But they are an inside-out version of art galleries. It is the outside walls that have to be guarded. How many guards do you need to station at the corners of a polygonal fortress in order to observe all of its exterior walls? There is a simple answer: the smallest whole number at least equal to $\frac{1}{2} \times$ (Number of corners). So, with 11 corners, you need 6 guards to guard the outside walls. Better still, we know it is the exact number you need. No fewer will be sufficient and no more are necessary. In the inside gallery problem, we only ever knew the largest possible number that we might need.

We can think about the case of right-angled prison walls again, as we did for the galleries. We might have an outer prison wall like these two right-angled shapes.

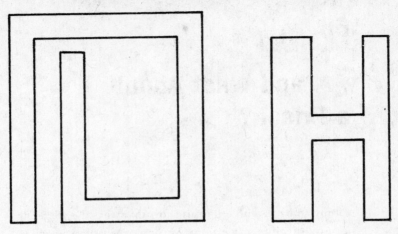

In these right-angled cases we need 1 plus the smallest whole number at least equal to ¼ × (Number of corners) to guard the whole of the outer wall. No fewer are possible and no more are needed. In the two examples shown there are 12 corners and so we need 1 + 3 = 4 guards.

67

A Snooker Trick Shot

Steve is going for the pink ball – and for those of you watching in black and white, the pink is next to the green.

Ted Lowe

Some people used to satisfy themselves that if their children spent most of their waking lives playing computer games, it was good for their understanding of maths and computing. I always wondered if they felt that hours at the snooker or pool hall were adding to their knowledge of Newtonian mechanics. Still, a knowledge of simple geometry can certainly provide you with some impressive snooker demonstrations for the uninitiated.

Suppose that you want to hit a single ball so that it goes around the table, bounces off three cushions, and returns to the point from which it was struck. Let's start with the easy case – a square table. Everything is nice and symmetrical, and it should be clear that you should place the ball in the middle of one of the sides of the table and then strike it at an angle of 45 degrees to the side. It will hit the middle of the adjacent side at the same angle and follow a path that is a perfect square, shown dashed in the diagram overleaf.

You don't have to start with the ball against one of the cushions, of course; if you strike it from any point on the dashed square path and send it along one of the sides of the dashed square, then the ball will eventually return to where it started (as long as you

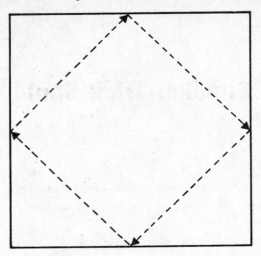

hit it hard enough). If you want it to stop at exactly the point that you hit it from, then you need to be very skilful – or at least put in a little practice.

Unfortunately, you won't very often come across a square snooker table. Modern tables are made from two squares side by side and a full-size table will have dimensions of 3.5 m × 1.75 m. The important general feature is that the table's length is twice its width. Using this simple fact, you can draw how your trick shot will have to run when the table has these rectangular dimensions. I have drawn in the diagonal as well for reference. Your shot has to go parallel to the diagonal and hit the sides at points that divide each of them in the ratio 2 to 1, the same ratio as the ratio of the length to the width of the table. (In the case of the square table this ratio was equal to 1 and you had to hit the centre of each side of the table.) This means that the angle that the ball's path makes with the long sides of the table has its tangent equal to ½, or 26.57 degrees, and the angle it makes with the short sides is 90 minus this angle, or 63.43 degrees, since the three interior angles of the right-angled triangles must add up to 180 degrees. The dashed parallelogram marks the only path on the rectangular table that will see the ball return to its starting point.

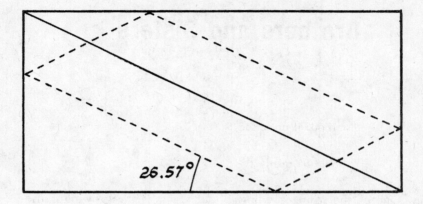

If you find yourself playing on a non-standard table, you need to recalculate. In general, the magic angle you need to strike the ball at, relative to the long side of the table, is the angle whose tangent equals the ratio of the width to the length of the table (1:2 for our full-size table and 1:1 for the square table) and the point against the cushion you need to strike from must divide the length of the side in the same ratio as the length to the width of the table.

68

Brothers and Sisters

Sisterhood is powerful.

Robin Morgan

One of the strangest things about China is the growing conse-quences of the 'one-child' policy. With the exception of multiple births (typically 1 per cent of the total), every young person in urban areas is an only child.* In this situation, the likelihood of each child being lavished with rather too much parental attention has given rise to the term 'Little Emperor Syndrome'. In the future, for almost everybody, there will be no brothers and sisters and no aunts or uncles. A concept like 'fraternity' will gradually lose all meaning.

At first, there seems to be a strange general asymmetry about brothers and sisters. If there are 2 children, one boy and one girl, then the boy has a sister but the girl doesn't. If there are 4 chil-dren, 3 girls and one boy, then the boy has got 3 sisters and the girls between them have $3 \times 2 = 6$ sisters. Each girl only gets to count the other girls as sisters but the boy counts them all. So it looks as if boys should always have more sisters than girls!

This seems paradoxical. Let's look more closely. If a family with n children has g girls and n−g boys, then the boys have a total of

---

* In rural areas a second child is permitted after an interval of 3 years if the first child was disabled or female.

g(n−g) sisters between them, while the girls have a total of $g(g-1)$ sisters. These numbers can only be equal if $g = \frac{1}{2}(n+1)$. This can never be true when n is an even number because then g would be a fraction.

The puzzle has been created because there are many ways in which a family of children can be formed. A family of 3 children can have 3 sons, 3 daughters, 2 sons and 1 daughter, or 2 daughters and 1 son. If we assume that there is an equal chance of $\frac{1}{2}$ that a newborn child will be a boy or a girl (this is not quite true in reality) then a family of n children can be made up in 2^n different ways. The number of different family make-ups with n children and g daughters is denoted by* ^nC_g and the boys will each have g(n−g) sisters. Given all the 2^n different ways in which the family can be divided into sons and daughters, we should be asking what is the *average* number of sisters that the boys in the n-child families will have. This is the sum of all the answers for the number of sisters they can have for all the possible values of the number g = 0,1,2, . . ., n divided by the total number, 2^n. This is

$$b_n = 2^{-n} \Sigma_g \; ^nC_g \times g(n-g)$$

Similarly, the average number of sisters that the girls in the n-child families have is

$$g_n = 2^{-n} \Sigma_g \; ^nC_g \times g(g-1)$$

The solutions to these formulae are much simpler than the formulae would lead us to expect. Remarkably, the average numbers of sisters for the boys and the girls are equal, and $b_n = g_n = \frac{1}{4} n(n-1)$. Notice that because it is an average, it doesn't mean that any given family has to have the average behaviour. When

---

* ^nC_g is shorthand for $n!/g!(n-g)!$ and is the number of ways of choosing g outcomes from a total of n possibilities.

n = 3, the average number of sisters is 1.5 (which no one family could have). When n = 4, the average number is 3. When n gets big, the number gets closer and closer to the square of n/2. Here is the table of the 8 possibilities for families with 3 children:

Family make-up with 3 children	Number of ways to make this family	Number of boys' sisters	Number of girls' sisters
3 boys	1	0	0
2 boys + 1 girl	3	2	0
2 girls + 1 boy	3	2	2
3 girls	1	0	6

We see that the total number of sisters for the boys are the sums from the second and third rows, $3 \times 2 + 3 \times 2 = 12$, and the total number of sisters for the girls is the sum of the third and the fourth rows; it is also $12 = 3 \times 2 + 1 \times 6$. Since there are 8 possible ways to make this family, the average number of sisters for the girls and the boys is equal to $12/8 = 1.5$, as predicted by our formula $\frac{1}{4} \times n \times (n-1)$ for the case of n = 3, when there are three children.

69

Playing Fair with a Biased Coin

And somewhat surprisingly, Cambridge have won the toss.

Harry Carpenter

Sometimes you need a fair coin to toss so that you can make a choice between two options without any bias. At the outset of many sports events, the referee tosses a coin and asks one of the competitors to call 'heads' or 'tails'. You could set up gambling games by scoring sequences of coin tosses. You could use more than one coin simultaneously, so as to create a greater number of possible outcomes. Now suppose that the only coin that you have available is a biased one: it does not have an equal probability (of ½) of falling 'heads' or 'tails'. Or perhaps you just suspect that the coin that your opponent has so thoughtfully provided might not be fair. Is there anything you can do in order to ensure that tossing a biased coin creates two equally likely, unbiased outcomes?

Suppose that you toss the coin twice and ignore outcomes where both outcomes are the same – that is, toss again if the sequence 'heads-heads' (HH) or 'tails-tails' (TT) happens. There are two pairs of outcomes that could result: 'heads' followed by 'tails' (HT), or 'tails' followed by 'heads' (TH). If the probability of the biased coin coming down 'heads' is p, then the probability of getting 'tails' is $1-p$, and so the probability of getting the sequence HT is p(1−p) and that of getting TH is (1−p)p. These two probabilities are the same, regardless of the probability p of the biased coins. All we have to do to get a fair game is define HEADS by the sequence HT and TAILS by the sequence TH, and the probability of TAILS is the same as the probability of HEADS. And you don't need to know the bias, p, of the coin.*

---

* This trick was thought up by the great mathematician, physicist and computer pioneer, John von Neumann. It had wide use in the construction of computer algorithms. One of the questions that was subsequently addressed was whether there were more efficient ways of defining the new HEAD and TAIL states. The way we have done it wastes 'time' by having to discard all the HH and TT outcomes.

70

The Wonders of Tautology

A good guide to understanding events in the modern world is to assume the opposite of what Lord Rees-Mogg tells you is the case.

Richard Ingrams

'Tautology' is a word that gives out bad vibes. It suggests meaninglessness, and my dictionary defines it as 'a needless repetition of an idea, statement or word'. It is a statement that is true in all eventualities: all red dogs are dogs. But it would be wrong to think that tautologies are of no use. In some sense they may be the only way to really secure knowledge. Here is a situation where your life depends upon finding one.

Imagine you are locked in a cell that has two doors – a red door and a black door. One of those doors – the red one – leads to certain death and the other – the black one – leads to safety, but you don't know which leads to which. Each door has a phone next to it from which you can call an adviser who will tell you which door you should take in order to get out safely. The trouble is that one adviser always tells the truth and the other always lies, but you don't know which one you are talking to. You can ask one question. What question should you ask?

Take the simplest question you could ask: 'Which door should

I go through?' The truthful adviser will tell you to go through the black one and the untruthful adviser will tell you to go through the red one. But, since you don't know which of the advisers you are speaking to is telling the truth, this doesn't help you. You would do just as well taking a random guess at red or black. 'Which door should I go through?' is not therefore a tautology in this situation. It is a question that can have different answers.

Suppose instead you asked: 'What door would the other adviser tell me to go through?' The situation is now more interesting. The truthful adviser knows that the lying adviser will tell you to go through the deadly red door so the truthful adviser will say that the other adviser will tell you to go through the red door. The untruthful adviser knows that the truthful adviser will tell you to go through the black door to safety and so the untruthful adviser will try to deceive you by saying you should go through the red door.

You have made a life-saving discovery. Regardless of who answers your question, you get told to go through the red door. You have stumbled upon a question that is a tautology – it always has the same answer – and that is your lifeline. Your strategy for safe escape is therefore clear: ask: 'What door would the other adviser tell me to go through?', note the answer (red) and then go through the other door (black) to safety.

71

What a Racket

Speed has never killed anyone, suddenly becoming stationary
. . . that's what gets you.

Jeremy Clarkson

Some things are harder than others to move. Most people think
that the only thing that counts in such a problem is their weight.
The heavier the load, the harder it is to shift it. But try moving
lots of different types of load and you will soon discover that the
concentration of the load plays a significant role. The more concentrated the mass, the easier it is to move and the faster it wobbles
(recall what we learnt in Chapter 2). Look at an ice skater beginning a spin. She will begin with her arms outwards and then steadily
draw them in towards her body. This results in an increasingly
rapid spin rate. As the skater's mass is more concentrated towards
her centre, she moves faster. On the other hand, if you look at
the shapes of girders used to build robust buildings, they have an
H-shaped cross-section. This distributes more mass away from the
centre of the girder and so makes it harder to move or deform
the girder when it is stressed.

This resistance to being moved is called 'inertia', following its
common usage, and it is determined by the total mass of an object
and also by its distribution, which will be determined by the shape
of the object. If we think still about rotation, then an interesting
example is a simple object like a tennis racket. It has an unusual

shape and can be rotated in three distinct ways. You can lay the tennis racket flat on the floor and spin it around its centre. You can stand it on its top and twist the handle. And you can hold it by the racket handle and throw it up in the air so it somersaults and returns to be caught by the handle again. There are three ways to spin it because there are three directions to space, each at right angles to the other two, and the racket can be spun around an axis pointing along any one of them. The racket behaves rather differently when spun around each of the different axes because its mass distribution is different around each axis and so it has a different inertia for motion around each axis. Here are two of them:

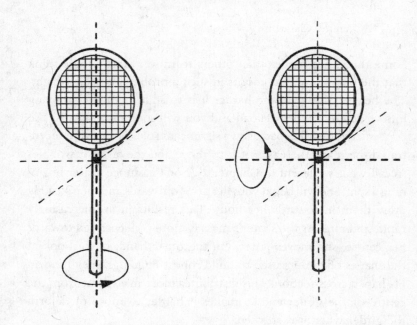

There is one remarkable property of these different motions, which is displayed by your three tosses of the racket. The motion around the axes about which the inertia is the largest or the smallest is simple. When the racket is flat on the ground or stood upright and spun, it does nothing very unusual. But when you spin it about the in-between axis, about which the inertia is intermediate between the largest and the smallest (shown on the right), something unusual happens. Hold the racket by the handle with the face upwards, as if you were holding a frying pan. Mark the top face with some chalk. Toss the racket so that it does a complete 360-degree turnover and catch it by the handle again. The face with the chalk mark will now be facing *downwards*.

The golden rule is that a spin around the axis with the in-between inertia is unstable. The slightest deviation from the precise centre line always causes it to flip over. Sometimes this is a good thing. If you are a gymnast doing a somersault on the beam, then you look more impressive (and score higher marks) if you do a twist as well. But the twist can happen automatically because of this instability.

A more serious example of this instability arose a few years ago on the International Space Station after it had been hit during a mistimed docking operation with a Russian supply craft. The Station was damaged and set in a slow spin. There was still gas in the system of retro rockets, which could be fired to slow the spin and return the Station to its usual state of equilibrium. The problem was, though, how should the rockets be fired? In what direction should you move the Station so as to counter the existing rotation. British astronaut Michael Foales had to solve this problem while holed up in the undamaged part of the Station with his laptop and a link to the ground. The most important things to discover were the three inertias of the Space Station with respect to rotations about its three axes. If the correcting rockets were fired wrongly, then they could set the Station spinning around its intermediate axis of inertia. The result would have been total disaster. The instability that flipped your tennis racket over had no bad effects on the racket, but flip over the Space Station and it would break apart, leaving all the astronauts dead, a quarter of a million kilograms of potentially lethal debris scattered in space and an astronomical financial loss. NASA didn't know the three inertias for the Space Station – no one had thought such facts would be needed – and so Foales had to work them out from the design plans and then calculate how the Station would respond to rockets being fired in different directions in order to correct its spin from the accident. Fortunately, he knew about the instability of rotation about the intermediate axis and got all his calculations right. The dangerous spin was righted and the astronauts were saved. Maths can be a matter of life and death.

72

Packing Your Stuff

On travel: I have learnt that you need four times as much water, twice as much money and half as many clothes as you think you need at the outset.

Gavin Esler

A young boy was once shown a large empty glass jar with a screw-top lid. He was handed a box of tennis balls and asked to fill the jar. He poured in some tennis balls and then moved them around a bit to try to squeeze in another tennis ball before screwing down the lid. 'Is the jar full?' he was asked. 'Yes, it's full,' he replied. But then he was given a box of marbles and asked to see if he could fit any more in the jar. He opened the lid and found that he could fit quite a few marbles in between the tennis balls. Giving the jar a shake now and then allowed the marbles to settle into the spaces. Eventually, he couldn't fit in another marble and announced that the jar was now full. His mentor then produced a bag of sand and asked the boy to fill the jar. Again, he unscrewed the lid and poured the sand into the top of the jar. This time he didn't need to fiddle around very much at all, just give the jar a careful shake now and again to make sure that the sand was pouring into all the empty nooks and crannies between the tennis balls and the marbles. Finally, he couldn't fit any more sand in the jar and screwed the lid back down again. The jar really was full!

There are some lessons to be learned from this story. If the boy had been given the sand first and asked to fill up the jar, then there would not have been any room to fit in any marbles or tennis balls afterwards. You need to start with the biggest things if there is to be any room for them at all. The same applies to more familiar packing problems. If you need to move lots of packages into a van then you might want to know how you should set about loading them in order to have the best chance of getting them all in. Our little story shows why you should start with the largest objects and then pack the next largest and so on, leaving the smallest until last.

The shapes of the things you are trying to pack clearly matter. Often, they are all the same size. If you are a manufacturer of sweets or other small food items, you might want to know what shape they should be in order to fit as many as possible into a jar or some other large storage container. For a long time it was thought that the answer was to make them all little spheres, like gobstoppers. Lots of little spheres seemed to give the smallest amount of empty space in between the closely packed balls. Interestingly, it turned out that this wasn't the best shape to use. If sweets are made in the shape of little ellipsoids, rather like mini-rugby balls or almonds, then more of the space can be filled by them. So Smarties and M&Ms fill a volume more efficiently than any collection of identical spheres. If the ellipsoids have their short to long axes in the ratio of 1 to 2, then they leave just 32% of the space empty, compared with 36% if they were made into spheres. This seemingly trivial fact has many important consequences for business efficiency and industrial manufacture, reducing wastage, shipping costs and the avoidance of unnecessary packaging.

73

Sent Packing Again

All my bags are packed; I'm ready to go.

John Denver

Our little packing problem with the jars in the previous chapter was a nice simple one. We started with the biggest objects and progressed down to the smaller ones. In practice, our problem may be more tricky. We may have lots of containers to fill up with shopping items of different sizes. How do we go about distributing the items of different sizes across the bags so as to use the smallest number of bags? 'Packing' might not just mean packing in space either; it can mean packing in time. Suppose you are the manager of a big copy-shop business that runs off copies of different documents of different sizes all day for customers. How do you allocate the different copying jobs to the machines so that you minimise the total number of machines that you need to complete the day's work?

These are all versions of a problem that computers find very time consuming to solve when the number of items to be packed and the number of 'boxes' in which to pack them becomes large.

Imagine that you can use large storage boxes that hold a maximum of 10 packages and you are given 25 packages of different sizes to stack in the boxes so as to use the smallest number of containers in the most efficient way. The sizes of the individual packages are as listed here:

6,6,5,5,5,5,4,3,2,2,3,7,6,5,4,3,2,2,4,4,5,8,2,7,1

First, let's imagine that these packages arrive on a conveyor belt so you can't sort them out as a group: you just stack them away one by one, as they come. The easiest strategy to follow is to put them into the first bin until the next won't fit, and then start a new bin. You are not allowed to go back and fill empty spaces in earlier bins because they have been taken away. The strategy is sometimes called 'Next Fit'. Here is how you end up filling the bins, starting from the left and adding a new one, as required. The 6-pack goes in the first box. The next 6-pack won't fit so we start a 2nd box. The 5-pack won't fit in there as well, so we start on a 3rd box. Adding the next 5-pack fills it and the next two 5-packs fill the next box and so on. Here is what ends up in the boxes if we follow this Next Fit prescription for the packages in the order given above :

[6], [6], [5,5], [5,5], [4,3,2], [2,3], [7], [6], [5,4], [3,2,2], [4,4], [5], [8,2], [7,1]

We have used 14 boxes and only three of them are full (the two [5,5]s and the [8,2]). The total amount of empty space in the unfilled boxes is $4+4+1+5+3+4+1+3+2+5+2 = 34$.

The large amount of wasted space has been caused by the fact that we can't go back and fill an empty space in an earlier box. How much better could you do if you could operate a packing strategy that allowed you to put packages in the first available box that had room? This is sometimes called 'First Fit' packing. Using First Fit, we start as before with a 6 and 6 in separate boxes then fill two boxes with two 5-packs. But the next is a 4-pack, which we can put in the first box with the 6. Next there is a 3-pack, which can fit in the second box with the 6-pack; then we can get two 2-packs and a 3-pack in the fifth box and so on, until we end by dropping the 1-pack back in the second box, which fills it. This is what the final packing allocation looks like:

[6,4], [6,3,1],[5,5],[5,5],[2,2,3,3],[7,2],[6,4],[5,2,2],[4,4],[5],[8],[7]

First Fit has done much better than Next Fit. We have only used 12 boxes, and the amount of wasted space is down to $1+1+2+5+2+3 = 14$. We have managed to fill six of the boxes completely.

We can now start to see how we might do better still in this packing business. The wasted space tends to come when we end up with a big package later on. The spaces left in the earlier boxes are all small ones by then, and we have to start a new box for each new package. We could clearly do better if we could do some sorting of the packages into descending size order. This may not always be an option if you are handling luggage coming off an airline's luggage belt, but let's see how much it helps when you can do it.

If we sort our packages by descending order of size then we end up with the new list:

8,7,7,6,6,6,5,5,5,5,5,5,4,4,4,4,3,3,3,2,2,2,2,2,1

Now let's try our old Next Fit strategy again with the sorting done – call it 'Sorted Next Fit'. The first six packages all go in their own boxes, then we fill three boxes with pairs of 5-packs, and so on. The end result looks like this:

[8],[7],[7],[6],[6],[6],[5,5],[5,5],[5,5],[4,4],[4,4],[3,3,3],[2,2,2,2,2],[1]

Bad luck at the end! We had to use a new box just to accommodate that last 1-pack. Using Sorted Next Fit we have ended up needing 14 boxes again – just as we had to do without the sorting – and we have wastage of 34 again. This is just the same as with the unsorted Next Fit. But if that final 1-pack hadn't been included we would have still needed 14 boxes for Next Fit but only 13 for Sorted Next Fit.

Finally, let's see what happens with a 'Sorted First Fit' strategy. Again, the first six packages go into separate boxes, the six 5-packs then fill three more boxes, but then the sorting comes into its own. Three of the 4-packs fill the boxes with the 6-packs in, while the other 4-pack starts a new box. The remaining packages fill the gaps nicely, leaving only the last box unfilled:

$$[8,2],[7,3],[7,3],[6,4],[6,4],[6,4],[5,5],[5,5],[5,5],[4,3,2,1],[2,2,2]$$

We have used 11 boxes, and the only wasted space is 4 in the last box. This is far superior to the performance of our other strategies and we could ask whether it is the best possible result. Could there be another packing strategy that used fewer boxes than 11? It is easy to see that there couldn't. The total size of all the packages we have is $1 \times 8 + 2 \times 7 + 3 \times 6 + \ldots + 5 \times 2 + 1 \times 1 = 106$. Since each box can only contain packages of total size 10, all the packages require at least $106/10 = 10.6$ boxes to accommodate them. So we can never use fewer than 11 boxes, and there will always be no fewer than at least 4 waste spaces.

In this case, we have found the best possible solution by using the Sorted First Fit method. If you look back at the very simple problem we looked at in the last section, fitting objects of three sizes into a jar, we were really using the Sorted First Fit strategy because we put the bigger objects in before the smaller ones. Unfortunately, packing problems are not all this easy. In general, there is no quick method for a computer to find the best way of packing any given assortment of packages into the smallest number of boxes. As the sizes of the boxes get bigger and the variety of sizes of the packages grows, the problem becomes computationally very hard and will eventually, if the number of packages gets big enough and their sizes diverse enough, defeat any computer to find the best allocation in a given time. Even in this problem, there are other considerations that might render the Sorted First Fit only second best. The sorting of the packages, on which the efficiency of the method depends, takes time. If the time taken to box up the packages is also a consideration, then just using fewer boxes may not be the most cost-effective solution.

74

Crouching Tiger

Tyger! Tyger! burning bright
In the forests of the night.

William Blake

A little while ago a tragic incident occurred at San Francisco Zoo, when Tatiana, a small (!) 135-kilogram Siberian tigress jumped over the wall of its enclosure, killed one visitor, and left two others seriously injured. Press reports quoted zoo officials as being amazed by the evidence that the tiger had managed to leap over the high wall around its enclosure: 'She had to have jumped. How she was able to jump that high is amazing to me,' said the zoo's Director, Manuel Mollinedo. Although at first it was claimed that the wall around the tiger enclosure was 5.5 metres high, it later turned out to be only 3.8 metres, a good deal lower than the 5 metres recommended for safety by the American Association of Zoos and Aquariums. But should we feel safe with any of these numbers? Just how high could a tiger jump?

The wall was surrounded by a dry moat that was 10 metres wide, and so the captive tiger faced the challenge of high-jumping 3.8 metres from a running start at a horizontal distance of at least 10 metres away from the wall at take-off. Over short distances on the flat a tiger can reach top speeds of more than 22 metres per second (i.e. 50 miles per hour). From a 5-metre start, it can easily reach a launch speed of 14 metres per second.

3.8m

10m

The problem is just that of launching a projectile. It follows a parabolic path up to its maximum height and then descends. The minimum
launch speed V that will achieve the vertical height h from a launch
point at a distance x in front of the wall is given by the formula

$$V^2 = g\,(h + \sqrt{(h^2 + x^2)})$$

Where $g = 10$ m/s^2 is the acceleration due to Earth's gravity.
Notice some features of the equation that show that it makes
sense: if gravity were made stronger (g gets bigger) then it is
harder to jump, and the minimum launch-speed, V, must get
bigger. Similarly, as the height of the wall, h, gets bigger or the
distance away at take-off, x, increases, a larger launch-speed is
needed to clear the wall.

Let's look at the set-up of the San Francisco Zoo's tiger enclosure shown above. The wall was 3.8 metres high, but the tiger is
bulky and its centre would have to clear* about 4.3 metres to get
cleanly over the wall because Siberian tigers are about 1 metre
high at the shoulder. (We will neglect the possibility that it clings
on and scrambles over the wall – always likely though.) This gives
$V^2 = 9.8(4.3 + \sqrt{(18.5 + 100)}) = 148.97$ (m/s)^2, so V is 12.2 m/s.

---

* In these problems the projectile is taken to be a mass of negligible size located
at its centre (a so-called 'point' mass). Of course, a tiger has a very significant
size and is in no sense a point. However, we shall neglect this and treat the tiger
as if it had its whole mass located at its centre.

This is within the launch-speed capability of a tiger, and so it could indeed have cleared the wall. Raise the wall to 5.5 metres so the tiger needs to raise its centre by 6 metres to clear the wall and the tiger needs to launch at 13.2 metres per second to clear the wall. As the Director said, 'Obviously now that something's happened, we're going to be revisiting the actual height.'

75

How the Leopard Got
His Spots

Then the Ethiopian put his five fingers close together . . . and
pressed them all over the Leopard, and wherever the five
fingers touched they left five little black marks, all close
together . . . Sometimes the fingers slipped and the marks got
a little blurred; but if you look closely at any Leopard now you
will see that there are always five spots – off five black fingertips.

Rudyard Kipling, 'How the Leopard Got His Spots'

Animal markings, particularly on big cats, are some of the most
spectacular things we see in the living world. These patterns are
by no means random, nor are they determined solely by the need
for camouflage. The activators and inhibitors that encourage or
block the presence of particular pigments flow through the embry-
onic animal in accord with a simple law that dictates how their
concentrations at different points depend on the amount of production of that pigment by chemical reactions and the rate at which
it spreads through the skin. The result is a wavelike spread of
signals, which will activate or suppress different colour pigments.
The resulting effects depend on several things, like the size and
shape of the animal and the wavelength of the pattern waves. If
you were to look at a large area of skin surface, then the peaks

and troughs of these waves would create a regular network of hills and valleys of different colours. The appearance of a peak occurs at the expense of the inhibiting tendency, and so you get a pronounced stripe or spot against a contrasting background. If there is a maximum concentration that is possible in a particular place, then the build-up of concentration will eventually have to spread out, and spots will merge, turning into patches or stripes.

The size of the animal is important. A very small animal will not have room for many ups and downs of the pigment-activating wave to fit along and around its body, so it will be one colour, or perhaps manage to be piebald like a hamster. When the animal is huge, like an elephant, the number of ups and downs of the waves is so enormous that the overall effect is monochrome. In between the big and the small, there is much more scope for variety, both from one animal to the next and over an animal's body. A cheetah, for example, has a spotty body but a stripy tail. The waves create separate peaks and troughs as they spread around the large and roughly cylindrical body of the cheetah, but when they spread to the thin cylindrical tail they were much closer together and merged to create the appearance of stripes. This tendency gives a very interesting mathematical 'theorem' that follows from the behaviour of the colour concentration waves on animals bodies: animals with spots can have striped tails but striped animals cannot have spotted tails.

76

The Madness of Crowds

The future belongs to crowds.

Don Delillo, *Mao II*

If you have ever been in a huge crowd, at a sports event, a pop concert or a demonstration, then you may have experienced or witnessed some of the strange features of people's collective behaviour. The crowd is not being organised as a whole. Everyone responds to what is going on next to them, but nonetheless the crowd can suddenly change its behaviour over a large area, with disastrous results. A gentle plodding procession can turn into a panic-stricken crush with people trying to move in all directions. Understanding these dynamics is important. If a fire breaks out or an explosion occurs near a large crowd, how will people behave? What sort of escape routes and general exits should be planned in large stadiums? How should religious pilgrimages of millions of worshippers to Mecca be organised so as to avoid repeating the deaths of hundreds of pilgrims that have occurred in the past, as panic generates a human stampede in response to overcrowding?

One of the interesting insights that informs recent studies of crowd behaviour and control is the analogy between the flow of a crowd and the flow of a liquid. At first one might think that understanding crowds of different people, all with different potential responses to a situation, and different ages and degrees of understanding of the situation, would be a hopeless task, but surprisingly, this is not the case. People are more alike than we might imagine. Simple local choices can quickly result in an overall order in a crowded situation. When you arrive at one of London's big rail termini and head down to the Underground system, you will find that people descending will have chosen the left- (or right-) hand stair, while those ascending will keep to the other one. Along the corridors to the ticket barriers the crowd will organise itself into two separate streams moving in opposite directions. Nobody planned all that or put up notices demanding it: it arose as a result of individuals taking their cue from what they observed in their close vicinity. This means that they act in response to how people move in their close vicinity and how crowded it is getting. Responses to the second factor depend a lot on who you are. If you are a Japanese manager used to travelling through the rush hour on the Tokyo train system, you will respond very differently to a crush of people around you than if you are a tourist visitor from the Scottish Isles or a school group from Rome. If you are minding young or old relatives, then you will move in a different way, linked to them and watching where they are. All these variables can be taught to computers that are then able to simulate what will happen when crowds congregate in different sorts of space and how they will react to the development of new pressures.

Crowds seem to have three phases of behaviour, just like a flowing liquid. When the crowding is not very great and the movement of the crowd is steady in one direction – like the crowd leaving Wembley Stadium for Wembley Park Underground station after a football match – it behaves like a smooth flow of a liquid. The crowd keeps moving at about the same speed all the time and there is no stopping and starting.

However, if the density of people in the crowd grows significantly, they start pushing against one another and movement starts to occur in different directions. The overall movement becomes more staccato in character, with stop-go behaviour, rather like a succession of rolling waves. The gradual increase in the density of bodies will reduce the speed at which they can move forward, and there will be attempts to move sideways if people sense that things might move forwards faster that way. It is exactly the same psychology as cars swopping lanes in a dense, slow-moving traffic jam. In both cases it sends ripples through the jam, which cause some people to slow and some people to shift sideways to let you in. A succession of those staccato waves will run through the crowd. They are not in themselves necessarily dangerous, but they signal the possibility that something much more dangerous could suddenly happen.

The closer and closer packing of people in the crowd starts to make them behave in a much more chaotic fashion, like a flowing liquid becoming turbulent, as people try to move in any direction so as to find space. They push their neighbours and become more vigorous in their attempts to create some personal space. This increases the risk of people falling, becoming crushed together so closely that breathing is difficult or children becoming detached from their parents. These effects can start in different places in a big crowd and their effects will spread quickly. The situation rapidly snowballs out of control The fallers become obstacles over which others fall. Anyone with claustrophobia will panic very quickly and react even more violently to close neighbours. Unless some type of organised intervention occurs to separate parts of the crowd from other parts and reduce the density of people, a disaster is now imminent.

The transition from smooth pedestrian flow to staccato movement and then crowd chaos can take anything from a few minutes to half an hour, depending on the size of the crowd. It is not possible to predict if and when a crisis is going to occur in a particular crowd, but, by monitoring the large-scale behaviour, the transition to the staccato movement can be spotted in different parts of a big crowd and steps taken to alleviate crowding at the key pressure points that are driving the transition where chaos will set in.

77

Diamond Geezer

I have always felt a gift diamond shines so much better than one you buy for yourself.

Mae West

Diamonds are very remarkable pieces of carbon. They are the hardest naturally occurring materials.

The most sparkling properties of diamonds, however, are optical, because diamond has a huge refractive index of 2.4, compared with that of water (1.3) or glass (1.5). This means that light rays are bent (or 'refracted') by a very large angle when they pass through a diamond. More important still, light that is shone onto a diamond surface at an angle more than just 24 degrees from the vertical to the surface will be completely reflected and not pass through the diamond at all. This is a very small angle – for light shone through air on water this critical angle is about 48 degrees from the vertical, and for glass it is about 42 degrees.

Diamonds also spread colours in an extreme fashion. Ordinary white light is composed of a spectrum of red, orange, yellow, green, blue, indigo and violet light waves, which travel at different speeds through the diamond and get bent by different angles (red the least, violet the most) as white light passes through a transparent medium. Diamond produces a very large difference between the greatest and the least bending of the colours, called its 'dispersion', and this creates the remarkable 'fire' of changing colours when light passes through a well-cut diamond. No other gem stones have such a large dispersive power. The challenge presented to the jeweller is to cut a diamond so that it shines as brightly and colour-fully as possible in the light it reflects back into the eye of the beholder.

The cutting of diamonds is an ancient practice that has gone on for thousands of years, but there is one man who contributed more than anyone to our understanding of how best to cut a diamond and why. Marcel Tolkowsky (1899–1991) was born in Antwerp into a leading family of diamond cutters and dealers. He was a talented child and after graduating from college in Belgium was sent to Imperial College in London to study engineering.* While still a graduate student there, in 1919 he published a remark-able book entitled *Diamond Design*, which showed for the first time how the study of the reflection and refraction of light within a diamond can reveal how best to cut it so as to achieve maximum brilliance and 'fire'. Tolkowsky's elegant analysis of the paths that are followed by light rays inside a diamond led him to propose a new type of diamond cut, the 'Brilliant' or 'Ideal', which is now the favoured style for round diamonds. He considered the paths of light rays coming straight at the top flat surface of the diamond and asked for the angles at which the back of the diamond should be inclined so as to completely internally reflect the light at the first and second internal reflections. This will result in almost all the incoming light passing straight back out of the front of the diamond and produce the most brilliant appearance. In order to appear as brilliant as possible, the outgoing light rays should not suffer significant bending away from the vertical when they exit the diamond after their internal reflections. The three pictures over-leaf show the effects of too great, and too small, an angle of cut compared to an optimal one which avoids light-loss by refraction through the back faces and diminished back-reflection.

* His doctoral thesis was on the grinding and polishing of diamonds rather than on their appearance.

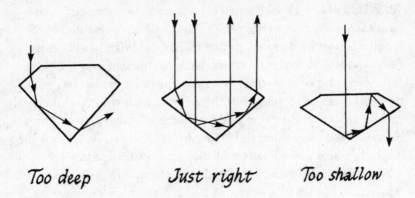

Too deep   Just right   Too shallow

Tolkowsky went on to consider the optimal balance between reflected brilliance and the dispersion of its spectrum of colours so as to create a special 'fire' and the best shapes for the different faces.[20]

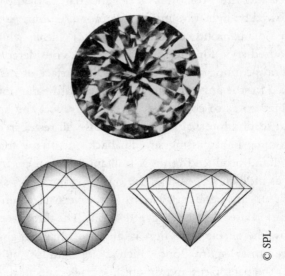

His analysis, using the simple mathematics of light rays, produced a recipe for a beautiful 'brilliant cut' diamond with 58 facets, and a set of special proportions and angles in the ranges needed to produce the most spectacular visual effects as a diamond is moved slightly in front of your eye. But you see there is more to it than meets the eye.

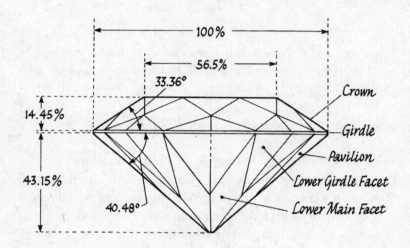

In the diagram we see the classic shape that Tolkowsky recommended for an ideal cut with the angles chosen in the narrow ranges that optimise 'fire' and brilliance. The proportions are shown for the parts of the diamond (shown with their special names) as percentages of the diameter of the girdle, which is the overall diameter.*

---

* The small thickness at the girdle is given so as to avoid a sharp edge.

78

The Three Laws of Robotics

For God doth know that in the day ye eat thereof, then your eyes shall be opened, and ye shall be as gods, knowing good and evil.

Book of Genesis

Yesterday I saw the film *I, Robot*, based on the robot stories of the great science-fiction writer Isaac Asimov. In 1942 he introduced the futuristic concept of humans coexisting with advanced robots in a short story entitled 'Runaround'. In order to ensure that humans were not destroyed or enslaved by their unerringly efficient assistants, he framed a set of 'Laws', which were programmed into the electronic brains of all robots as a safeguard. What those laws should be is an interesting question, not merely one of technological health and safety, but a deeper issue for anyone wondering why there is evil in the world and what steps a benevolent Deity might have taken to stop it.

Asimov's original three laws are modelled on the three laws of thermodynamics

First Law: A robot may not injure a human being or, through inaction, allow a human being to come to harm.

Second Law: A robot must obey orders given to it by human beings, except where such orders would conflict with the First Law.

Third Law: A robot must protect its own existence as long as such protection does not conflict with the First or Second Law.

Later, Asimov added the 'Zeroth Law', again as in thermodynamics, to stand before the First Law:

Zeroth Law: A robot may not harm humanity, or, by inaction, allow humanity to come to harm.

The reason for this last legal addition is not hard to find. Suppose a madman had gained access to a nuclear trigger that could destroy the world and only a robot could stop him from pressing it, then the First Law would prevent the robot from acting to save humanity. It is inaction on the part of robots that is a problem with the First Law, even when the Zeroth Law is irrelevant. If my robot and I are shipwrecked on a desert island and my gangrenous foot needs to be amputated in order to save my life, will my robot be able to overcome the First Law and cut it off? And could a robot ever act as a judge in the courts where he must hand down punishments to those found guilty by a jury?

Should we feel safe if robots were created in large numbers with these four laws programmed into their electronic brains? I think not. It's all a question of timing. The precedence of the Zeroth Law over the First means that the robot may kill you because you are driving a gas guzzling car or not recycling all your plastic bottles. It judges that your behaviour, if it continues, threatens humanity. It might become very concerned about its duty to act against some of the world's political leaders as well. Asking robots to act for the good of humanity is a dangerous request. It seeks something that is not defined. There is no unique answer to the question 'What is the good of humanity?' No computer could exist that prints out a list of all actions that are good for humanity and all actions that are harmful to it. No programme can tell us all good and all evil.

You might feel safer without the Zeroth Law than with it. Still, there is another worrying consideration that could put you at risk from all the harmful direct actions that the First, Second and Third Laws were conceived to protect us from. Advanced robots will have complicated thoughts, thoughts about themselves and us as well as about inanimate objects: they will have a psychology. Just as with humans, this may make them hard to understand – but it may also lead them to suffer from some of the psychological problems that humans can fall victim to. Just as it is not unknown for humans to be deluded into thinking they are robots, it may be that a robot could think it was a human. In that situation it could do what it likes because it no longer believes that the Four Laws of Robotics apply to it. Closely linked to this problem would be the evolution of religious or mystical beliefs in the robot mind. What then of the Third Law? What robotic existence is it that must be preserved? The material fabric of the robot? The soul in the machine that it perceives itself to have? Or, the 'idea' of the robot that lives on in the mind of its maker?

You can carry on asking questions like this for yourself, but you can see that it is not so easy to trammel the consequences of artificial intelligence by imposing constraints and rules on its programming. When that 'something' that we call 'consciousness' appears, its consequences are unpredictable, with enormous potential for good or evil, and it is hard to have one without the other – a bit like real life really.

79

Thinking Outside the Box

Many people would rather die than think; in fact, most do.

Bertrand Russell

It is easy to get trapped into thinking about a problem in one set way. Breaking out and being 'imaginative' or original in solving a problem can require a different way of thinking about it rather than just a correct implementation of principles already learned. Simple problems that involve the application of fixed rules in a faultless way can usually be conquered by the second approach. For example, if someone challenges you to a game of noughts and crosses you should never lose, regardless of whether you move first or second. There is a strategy whose worst outcome is a draw, but it will give you a win only if your opponent deviates from that optimal strategy. Alas, not all problems are as easy as finding the best move in noughts and crosses. Here is an example of a simple problem whose solution will almost certainly take you by surprise.

Write down a 3×3 square of nine dots. Now pick up your pencil and *without lifting the pencil point from the paper or retracing your path*, draw four straight lines that pass through all the dots.

Here's one failed attempt. It misses out one of the points in the middle on the left-hand edge:

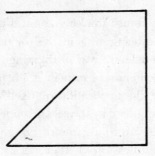

Here's another. It's also one point short because it doesn't go through the central point:

It looks impossible doesn't it. I can do it with four straight lines only if I retrace the pencil path, going down the diagonal and then going back and forth along the intersecting lines. But that requires the drawing of far more than four lines, even though only four appear to be present when you have finished:

There *is* a way to draw four lines through all the points without lifting the pencil or retracing its path, but it requires breaking a rule that you imposed on yourself for no reason at all. It wasn't one of the restrictions imposed at the outset. You were just so used to playing by a certain type of rule that you didn't think to step outside the box and break it. The solution that you want simply requires that you draw straight lines that (literally) extend beyond the box of nine points before they turn back in a different direction.

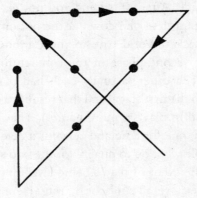

Think outside the box!

80

Googling in the Caribbean – The Power of the Matrix

Cricket is organised loafing.

William Temple

Most sports create league tables to see who is the best team after all the participants play each other. How the scores are assigned for wins, losses and draws can be crucial for determining who comes out on top. Some years ago, football leagues decided to give three points for a win rather than two, in the hope that it would encourage more attacking play. A team would get far more credit for winning than for playing out a draw – in which each team earns only one point. But somehow this simple system seems to be rather crude. After all, should you not get more credit for beating a top team than one down at the bottom of the league?

The 2007 Cricket World Cup in the Caribbean gives a nice example. In the second phase of the competition, the top eight teams played each other (actually each had played one of the others already in the first stage, and that result was carried forward so they only had to play six more games). They were given two points for a win, one for a tie and zero for a defeat. The top four teams in the table went on to qualify for the two semi-final knock-out games. In the event of teams being level on points they were separated by their overall run-scoring rate. Here is the table:

Super Eight Standings

TEAM	M	W	D	L	Net R/R	Points
Australia	7	7	0	0	2.40	14
Sri Lanka	7	5	0	2	1.48	10
New Zealand	7	5	0	2	0.25	10
South Africa	7	4	0	3	0.31	8
England	7	3	0	4	-0.39	6
West Indies	7	2	0	5	-0.57	4
Bangladesh	7	1	0	6	-1.51	2
Ireland	7	1	0	6	-1.73	2

But, let's think about another way of determining the team rankings that gives more credit for beating a good team than a bad one. We give each team a score that is equal to sum of the scores of the teams that they beat. Since there were no tied games we don't have to worry about them. The overall scores look like a list of eight equations:

$$A = SL + N + SA + E + W + B + I$$
$$SL = N + W + E + B + I$$
$$N = W + E + B + I + SA$$
$$SA = W + E + SL + I$$
$$E = W + B + I$$
$$W = B + I$$
$$B = SA$$
$$I = B$$

This list can be expressed as a matrix equation for the list $\underline{x} = (A, N, W, E, B, SL, I, SA)$ with the form $A\underline{x} = K\underline{x}$, where K is a constant and $\mathbf{A}$ is an 8×8 matrix of 0's and 1's denoting defeats and wins, respectively, and is given by:

	A	N	W	E	B	SL	I	SA
A	0	1	1	1	1	1	1	1
N	0	0	1	1	1	0	1	1
W	0	0	0	0	1	0	0	0
E	0	0	1	0	1	0	1	0
B	0	0	0	0	0	0	0	1
SL	0	1	1	1	1	0	1	0
I	0	0	1	0	1	0	0	0
SA	0	0	1	1	0	1	1	0

In order to solve the equations and find the total scores of each team, and hence their league ranking under this different point-scoring system, we have to find the eigenvector of the matrix **A** with all its entries positive or zero. Each of these solutions will require K to take a specific value. This corresponds to a solution for the list **x** in which all have positive (or zero – if they lost every game) scores, as is obviously required for the situation being described here. Solving the matrix for this, so-called 'first-rank' eigenvector, we find that it is given by

$$\underline{x} = (A, N, W, E, B, SL, I, SA) = (0.729, 0.375, 0.104, 0.151, 0.153, 0.394, 0.071, 0.332)$$

The ranking of the teams is given by the magnitudes of their scores here, with Australia (A) at the top with 0.729 and Ireland (I) at the bottom with 0.071. If we compare this ranking with the original table we have:

Super Eight standings	My Ranking
A	A 0.729
SL	SL 0.394
N	N 0.375
SA	SA 0.332

E		B 0.153
W		E 0.151
B		W 0.104
I		I 0.071

The top four teams qualifying for the semi-finals finish in exactly the same order under both systems, but three of the bottom four are quite different. Bangladesh won only one game, so scored a mere two points and finished second from bottom of the World Cup League. Under our system they jump up to fifth because their one win was against the higher ranked South Africans. England actually won two games but only against the bottom two teams, and end up ranked just behind Bangladesh (although it takes the third decimal place to separate them – 0.153 vs. 0.151). The poor West Indies finished sixth under the straightforward league system but drop a position under the rank system.

This system of ranking is what lies at the root of the Google search engine. The matrix of results when team i plays team j corresponds to the number of web links that exist between topic i and topic j. When you search for a term, a matrix of 'scores' is created by the massive computing power at Google's disposal, which solves the matrix equation to find the eigenvector, and hence the ranked list of 'hits' to the word that you were searching for. It still seems like magic though.

81

Loss Aversion

In theory there is no difference between theory and practice. In practice there is.

Yogi Berra

People seem to react very differently to the possibility of gains and losses. Economists took a long time to recognise that human behaviour is not symmetrical in this respect when it comes to decision making. We tend to be naturally risk averse and work much harder to avoid a small loss than to secure a larger gain. Being 'loss averse' means that losing a £50 note in the street gives you more unhappiness than the happiness you enjoy if you find a £50 note. You feel better about avoiding a 10 per cent surcharge than taking advantage of a 10 per cent reduction in train ticket prices.

Imagine that you are a market trader who sells goods from a roadside stall. You decide that you want to obtain a certain income each day and you will carry on working until you achieve that level of sales. What happens? When trade is good you quickly reach the sales target and go home early. When trade is bad you carry on working longer and longer hours in order to meet your target. This seems irrational. You work far longer in order to avoid a shortfall in your target, but you don't seize the opportunity to work longer when the demand is high. You are a classic example of the psychology of risk aversion.

Some people would argue that this type of behaviour is just

irrational. There is no good reason for it. On the other hand, gains and losses are not necessarily symmetrical with respect to the amount of money that you currently have. If your total wealth is £100,000, then a gain of £100,000 is to be welcomed, but a loss of £100,000 is to be avoided much more because it will bankrupt you. The potential loss is much greater than the possible gain.

Sometimes the taking of decisions does rest upon a purely psychological perception of apparent differences that do not truly exist. As an example, suppose the Environment Agency has to draw up plans to counter the effects on coastal homes of an anomalously high tide and expected storm surge that is expected to wreck 1,000 homes. It asks people to choose between two plans. Plan A uses all resources to build a wall in one location and will save 200 homes. Plan B uses the resources more diversely and will save all 1,000 homes from destruction with a probability* of 1/5. Faced with this choice, most people pick the sure and positive sounding Plan A.

Imagine, now, that the Environment Agency has a different Public Relations Officer who wants to present these two plans differently. The choice is now going to be between Plan C, which allows 800 homes to be destroyed, and Plan D, which leads to no homes being destroyed with a probability of 1/5 and all 1,000 homes being destroyed with a probability† of 4/5. Most people choose Plan D. This is strange because Plan D is the same as Plan B, and Plan A is the same as Plan C. Our innate risk aversion makes us pick D over C, but not B over A, because we are more sensitive to losses. The sure loss of 800 homes seems worse to us than the 1/5 chance of losing 1,000. But when it comes to the saving of homes, we don't respond so strongly to the chance of saving 1,000 as we do to the surety of saving 200. Odd.

---

* This means that the expected number of homes that will be saved is 1,000 × 1/5 = 200, the same number saved in Plan A.
† The expected number of homes destroyed is 800 in both Plans C and D, i.e. the expected number saved is 200, as in Plans A and B.

82

The Lead in Your Pencil

We are all pencils in the hand of God.

Mother Teresa

The modern pencil was invented in 1795 by Nicholas-Jacques Conte, a scientist serving in the army of Napoleon Bonaparte. The magic material that was so appropriate for the purpose was the form of pure carbon that we call graphite. It was first discovered in Europe, in Bavaria at the start of the fifteenth century, although the Aztecs had used it as a marker several hundred years earlier. Initially it was believed to be a form of lead and was called 'plumbago' or black lead (hence the 'plumbers' who mend our lead water-carrying pipes), a misnomer that still echoes in our talk of pencil 'leads'. It was called graphite only in 1789, using the Greek word 'graphein' meaning 'to write'. Pencil is an older word, derived from the Latin 'pencillus', meaning 'little tail', to describe the small ink brushes used for writing in the Middle Ages.

The purest deposits of lump graphite were found in Borrowdale near Keswick in the Lake District in 1564 and spawned quite a smuggling industry and associated black economy in the area. During the nineteenth century a major pencil manufacturing industry developed around Keswick in order to exploit the high quality of the graphite. The first factory opened in 1832, and the Cumberland Pencil Company has just celebrated its 175th anniversary, although the local mines have long been closed and supplies of the graphite used now come from Sri Lanka and other far away places. Cumberland pencils were those of the highest quality because the graphite used shed no dust and marked the paper very well. Conte's original process for manufacturing pencils involved roasting a mixture of water, clay and graphite in a kiln at $1,900^0$ Fahrenheit before encasing the resulting soft solid in a wooden surround. The shape of that surround can be square, polygonal or round, depending on the pencil's intended use – carpenters don't want round pencils that are going to roll off the workbench. The hardness or softness of the final pencil 'lead' can be determined by adjusting the relative fractions of clay and graphite in the roasting mixture. Commercial pencil manufacturers typically market 20 grades of pencil, from the softest, 9B, to the hardest 9H, with the most popular intermediate value, HB, lying midway between H and B. 'H' means hard and 'B' means black. The higher the B number, the more graphite gets left on the paper. There is also an 'F', or Fine point, which is a hard pencil for writing rather than drawing.

The strange thing about graphite is that it is a form of pure carbon that is one of the softest solids known, and one of the best lubricants because the six carbon atoms that link to form a ring can slide easily over adjacent rings. Yet, if the atomic structure is changed, there is another crystalline form of pure carbon, diamond, that is one of the hardest solids known.

An interesting question is to ask how long a straight line could be drawn with a typical HB pencil before the lead was exhausted. The thickness of graphite left on a sheet of paper by a soft 2B pencil is about 20 nanometres and a carbon atom has a diameter of 0.14 nanometres, so the pencil line is only about 143 atoms thick. The pencil lead is about 1 mm in radius and therefore π square mm in area. If the length of the pencil is 15 cm, then the volume of graphite to be spread out on a straight line is 150π cubic mm. If we draw a line of thickness 20 nanometres and width 2 mm, then there will be enough lead to continue for a distance $L = 150\pi/4 \times 10^{-7}$ mm $= 1,178$ kilometres. But I haven't tested this prediction!

83

Testing Spaghetti to Destruction

Every time I see a Parceline van I shall remember Miles
Kington. Because it was Miles who had decided that it was the
name of an Italian pasta dish.

Richard Ingrams

Hold both ends of a long, brittle, rod of dry spaghetti. Flex it and
gradually move the ends together so that the rod snaps. You might
have expected that eventually the rod would snap into two pieces,
leaving you holding one in each hand. Strangely, this never happens.
The spaghetti always breaks into more than two pieces. This is
odd. If you had snapped a thin rod of wood or plastic it would
have broken into two pieces. Why does the spaghetti behave differently? Richard Feynman was puzzled by this question as well and
a story appears in his biography, told by Daniel Hillis:

Once we were making spaghetti . . . If you get a spaghetti stick and you
break it, it will almost always break into three pieces. Why is this true – why
does it break into three pieces? We spent the next two hours coming up with
crazy theories. We thought up experiments, like breaking it underwater
because we thought that might dampen the sound, the vibrations. Well, we
ended up at the end of a couple of hours with broken spaghetti all over the
kitchen and no real good theory about why spaghetti breaks in three.

More recently some light has been shed on this problem, which turned out to be unexpectedly difficult. A brittle rod of anything, not just spaghetti, will break when it gets curved by more than a critical amount, called its 'rupture curvature'. There is no mystery about that, but what happens next is interesting. When the break first occurs, one end of each part will be left free while the other end is held in your hand. The free end that has suddenly been released tries to straighten itself and sends waves of curvature back along its length towards your hand where it is held fixed. These waves reflect and meet others arriving at different places along the spaghetti rod. When they meet, a sudden jump in curvature occurs, sufficient to break the flexed spaghetti again. New waves of curvature get produced by this new breaking and can lead to more local increases in curvature beyond the critical value at different points in the spaghetti. As a result, the spaghetti will break in one or more other places after it first fractures. The breaking stops when there is no longer enough energy left to allow the waves to travel along the pasta rod you are left holding. Any fragments that find themselves free at both ends just fall to the ground.

84

The Gherkin

Think cool; think cucumber.

Stephen Moss

The most dramatic modern construction in the City of London is 30 St Mary Axe, more commonly known as the Swiss Re building, the Pine Cone or simply the Gherkin. Prince Charles sees it as symptomatic of a rash of carbuncular towers on the face of London. The architects, Norman Foster and Partners, heralded it as a signature building for the modern age and received the 2004 RIBA Stirling Prize for their creation. It has succeeded in putting the Swiss Re insurance company in the public eye and has stimulated a wide-ranging debate about the desirability of towers on the traditional horizons and sight-lines of the City of London. Alas, while there is an ongoing debate about the aesthetic success of the Gherkin, there is not much doubt that it has been a bit of a commercial disappointment for Swiss Re. The company occupies just the first 15 of the 34 floors, but has never succeeded in renting the other half of the building to another, single, organisation. This is not entirely surprising: the type of high-profile commercial enterprise able to afford such space would recognise that the building has become so totally associated with the name of Swiss Re that it would be forever playing second fiddle and would gain no kudos at all by its presence there. As a result the space has been parcelled up into smaller lets.

The most obvious feature of the Gherkin is that it's big – 180 metres high – and the creation of a tower on such a scale creates structural and environmental problems. Today, engineers can create sophisticated computer models of a big building that enable them to study its response to wind and heat, its take-up of fresh air from the outside, and its effect on passers-by at ground level. Tinkering with one aspect of the design, like the reflectivity of its surface, will have effects in many other areas – changing the internal temperature and air-conditioning requirements, for instance – and all the consequences can be seen at once using sophisticated computer simulations of the building. It is no good following a 'one thing at a time' approach to designing a complicated structure like a modern building, you have to do a lot of things all at once.

The Gherkin's elegant curved profile is not just driven by aesthetics or some mad designer's desire to be spectacular and controversial. The tapering shape, starting narrowest at street level and bulging most at floor 16, before narrowing again steadily towards the top, was chosen in response to the computer models.

Tall buildings funnel winds into narrow channels around them at street level (it's just like putting your finger partly over the nozzle of a garden hose to make the jet reach further – the increased pressure brought about by the constriction results in a higher velocity of waterflow) and this can have a horrible effect on passers-by and people using the building. They feel as if they are in a wind tunnel. The narrowing of the building at the base reduces these unwanted wind effects because there is less constriction of the airflows. The tapering of the top half also plays an important role. If you stand at ground level beside a conventional untapered tower block and look upwards, the building dwarfs you and blots out a large fraction of the sky. A tapering design opens up more of the sky and reduces the dominating effect of the structure because you can't see the top from close-by on the ground.

The other striking feature of this building's exterior is that it is round, not square or rectangular. Again, this is advantageous for smoothing and slowing the airflow around the building. It also assists in making the building unusually eco-friendly. Six huge triangular wedges are cut into each floor level from the outside in. They bring light and natural ventilation deep into the heart of the building, reducing the need for so much conventional air-conditioning and making the building twice as energy efficient as a typical building of the same scale. These wedges are not set immediately below each other from floor to floor, but are slightly rotated with respect to those on the floors above and below. This helps increase the efficiency of the air suction into the interior. It is this slight offsetting of the six wedges from floor to floor that creates the exterior spiral pattern that is so noticeable.

Looking at the rounded exterior from a distance, you might have thought that the individual surface panels are curved – a complicated and expensive manufacturing prospect – but in fact they are not. The panels are small enough, compared with the distance over which the curvature is significant, that a mosaic of flat, four-sided panels is quite sufficient to do the job. The smaller you make them, the better they will be able to approximate the covering of the curved outside surface. All the changes of direction are made at the angles joining different panels.

85

Being Mean with the Price Index

The average human has one breast and one testicle.

Des McHale

All economically developed countries have some measure of the change of the average person's cost of living that is derived from the price of standard units of some collection of representative goods, including staple foods, milk, heat and light. They have names like the Retail Price Index (RPI) or the Consumer Price Index (CPI), and are a traditional measure of inflation, which can then be used to index salaries and benefits to allow for it. Citizens therefore want these measures to come out on the high side, whereas governments want them to be on the low side.

One way to work out a price index is to take a simple average of a collection of prices – just add up the prices and divide by the number of different prices you totalled. This is what statisticians call the *arithmetic mean*, or simply the 'average'. Typically, you want to see how things are changing over time – is the cost of the same basket of goods going up or down from month to month? – so you want to compare the average last year with its value this year by dividing the first by the second. If the result is bigger than 1, then prices are going down; if it is less than 1, then they are going up. This is simple enough, but are there hidden problems?

Suppose that a family habitually spends the same amount of money each week on beef and fish and then the price of beef doubles while fish costs the same. If they keep buying the same amount of beef and fish, their total bill for beef and fish will be 1.5 times their old one, an increase of 50%. The average of the price changes will be $\frac{1}{2} \times (1 + 2) = 1.5$. The $\frac{1}{2}$ is just dividing by the number of products (two: fish and beef); 1 is the factor by which the price of fish changes (it stays the same) and 2 is the factor by which the price of beef changes (it doubles).

This 1.5 inflation factor, an increase of 50 per cent, will be the headline statistic. But for a non-meat-eating family that eats no beef it will be meaningless. If they were only eating fish they will have seen no change in their weekly bill at all. The inflation factor is also an average over all possible family eating choices. It is based on an assumption about human psychology. It assumes that the family will go on eating the same amount of beef as fish, despite the rise in the relative cost of beef compared to fish. In reality, families might behave differently and decide to adjust the amount of fish and beef they buy, so that they still spend the same amount of money on each. This will mean they buy less beef in the future because of its increased cost.

The assumption that families respond to price changes by keeping constant the fraction of their budget that they spend on each commodity suggests that the simple arithmetic mean price index should be replaced by another sort of mean.

The *geometric mean* of two quantities is the square root of their product.* The geometric mean index of the price change in beef and no change in the price of fish is therefore:

$$\sqrt{(\text{New beef cost}/\text{old beef cost})} \times \sqrt{(\text{New fish cost}/\text{old fish cost})}$$
$$= \sqrt{1} \times \sqrt{2} = 1.41$$

---

* More generally, the geometric mean of n quantities is the nth root of their product.

The interesting thing about these two measures of inflation is that the geometric mean of any number of quantities is never greater than the arithmetic mean[21] so governments certainly prefer the geometric mean.* It suggests lower inflation and results in a lower inflationary index for wage increases and social security benefits.

Another benefit of the geometric mean is practical rather than political. To determine the inflation rate you compare the index at different times. If you were using the arithmetic mean then you would need to work out the ratio of the arithmetic means in 2008 and 2007 to find out how much the 'basket' of prices inflated in price over the last year. But the arithmetic mean involves the *sum* of all sorts of different things that can be measured in different units: £ per Kg, £ per litre, etc; some involve price per unit weight, some price per unit volume. It's a mixture, which is a problem because in order to calculate the arithmetic mean you have to add them together, and this doesn't make any sense if the units of the different commodities that go into it are different. By contrast, the beautiful feature of using the geometric mean is that you can use any units you like for the commodity prices in 2008 as long as you use the same ones for the same commodities in 2007. When you divide the geometric mean value for 2008 by its value for 2007, to find the inflation factor, all the different units cancel out because they are exactly the same in the denominator and in the numerator. So, it's a pretty mean index.

---

* In the United States the Labor Bureau changed the calculation of the Consumer Price Index from the arithmetic to the geometric mean in 1999.

86

Omniscience can be a Liability

For sale by owner, Encyclopaedia Britannica, excellent condition. No longer needed. Husband knows everything.

Small ad

Imagine what it must be like to know everything. Well, maybe it's not so easy. Perhaps it's more manageable to imagine what it would be like to know everything that you wanted to know – or needed to know. Even that sounds like being in a commanding position: you know next week's winning lottery numbers, which train to take so as to avoid delays, who is going to win the big football game. There are huge advantages to be had, although life could eventually prove miserable without the benefit of the odd pleasant surprise.

There is a strange paradox about knowing everything that shows that you can find yourself worse off than if you didn't know everything. Suppose you are watching a dare-devil game of 'chicken' in which two stunt pilots fly aircraft towards each other at high speed (it's like aerial jousting without the horses and lances). The loser is the flyer who first swerves off to the side. What should a pilot's tactic be in such a game? If he never swerves, then he will end up dead if the other flyer has the same tactic – no one wins then. If he always swerves he will never win – only draw when the other pilot swerves as well. Clearly, always swerving is the only sure strategy that minimises loss. Some mixed strategy of some-times swerving and sometimes not swerving will produce some wins, but ultimately will result in death unless the other pilot always swerves when the other does not. If the other pilot thinks the same way (he is thinking this through independently and doesn't know what his rival has decided), he should draw the same conclusions.

Now play this game with an omniscient opponent. He knows what your strategy is going to be. So you should·choose *never* to swerve. He will know that your strategy is never to swerve and will therefore choose to swerve every time. The omniscient pilot will never win!

This story presumably has an application to the world of espi-onage. If you are listening-in on all your enemies' communica-tions and they know that you are listening, then your omniscience may put you at a disadvantage.

87

Why People aren't Cleverer

God help in my search for truth, and protect me from those who believe they have found it.

Old English Prayer

When astronomers speculate about the nature of advanced extra-terrestrials or biologists contemplate a future where humans have evolved to be smarter than they are today, it is always assumed that increased intelligence has to be a good thing. The evolutionary process increases the likelihood of passing on traits that increase the chance of survival and having offspring. It is hard for us to imagine how it might become a liability to be more intelligent on the average as a species than we are.

If you have ever had experience of trying to manage a community of cleverer-than-average individuals, then you could easily be led to think otherwise. A good example might be the challenge faced by the chairperson of a university department or the editor of a book to which many authors are contributing. In such situations you soon realise that this type of high intelligence tends to accompany a tendency to be individualistic, to think independently and to disagree with others who think differently. Perhaps the ability to get on with others, to work with others rather than against them, was more important during the early evolution of intelligence. If intelligence now evolves rapidly to superhuman levels, perhaps the effects would be socially disastrous? Then again, low levels of average intelligence are clearly disadvantageous when it comes to dealing with potentially foreseeable hazards. There might well be an optimal level of intelligence for life in a given environment in order to maximise the chances of long-term survival.

88

The Man from Underground

Art has to move you and design does not, unless it's a good design for a bus.

David Hockney

Once I saw two tourists trying to find their way around central London streets using an Underground train map. While this is marginally better than using a Monopoly board, it is not going to be very helpful. The map of the London Underground is a wonderful piece of functional and artistic design, but it has one striking property: it does not place stations at geographically accurate positions. It is a *topological* map: it shows the links between stations accurately, but distorts their actual positions for aesthetic and practical regions.

When Harry Beck first proposed this type of map to the management of the London Underground Railway, he was a young draughtsman with a background in electronics. The Underground Railway was formed in 1906, but by the 1920s it was failing commercially, not least because of the apparent length and complexity of travelling from its outer reaches into central London, especially if changes of line were necessary. A geographically accurate map looked a mess because of the higgledy-piggledy nature of inner London's streets that had grown up over hundreds of years without any central planning. It was not New York, nor even Paris, with a simple overall street plan. People were put off.

Beck's elegant 1931 map – although initially turned down by the railway's publicity department and the Underground's managing director, Frank Pick – solved many problems at one go. Unlike any previous transport map, and reminiscent of an electronic circuit board, it used only vertical, horizontal and 45-degree lines. It also eventually drew in a symbolic River Thames, introduced a neat way of representing the exchange stations and distorted the geography of outer London to make remote places like Rickmansworth, Morden, Uxbridge and Cockfosters seem close to the heart of London. Beck continued to refine and extend this map over the next 40 years, accommodating new lines and extensions of old ones, always striving for simplicity and clarity. It was always referred to by him as the London Underground Diagram, or simply 'The Diagram', to avoid any confusion with traditional maps.

Beck's classic piece of design was the first topological map. It can be changed by stretching it and distorting it in any way that doesn't break the connections between stations. Imagine it drawn on a rubber sheet, which you then stretch and twist in any way you like without cutting or tearing it. Its impact was sociological as well as cartographical. It redefined how people regarded London. It drew in the outlying places on the map and made their residents feel that they were close to central London. It even defined the house price contours. For most of us, it is the picture of how London 'is'.

Beck's original idea makes good sense. When you are below ground on the Underground you don't need to know where you are, as you do if you are travelling on foot or by bus. All that matters is where you get on and off and how you change onto other lines. Pushing far away places in towards the centre doesn't only help Londoners feel more connected, it helps create a neat beautifully balanced diagram that fits nicely on a small fold-out sheet that will go into your jacket pocket.

89

There are No Uninteresting Numbers

Everything is beautiful in its own way.

Ray Stevens

There is no end to the list of numbers. The small ones, like 1, 2 and 3, are in constant use to describe the small numbers of life: the number of children, cars, items on a shopping list. The fact that there are so many words for groups of small quantities that are specific to their identities – for instance, double, twin, brace, pair, duo, couple, duet, twosome – suggests that their origins predate our decimal counting system. Each of these small numbers seems to be interesting in some way. The number 1 is the smallest of all, 2 is the first even number, 3 is the sum of the previous two, 4 is the first that is not prime and so can be divided by a number other than itself, 5 is the sum of a square (2^2) plus 1. And so we might go on. Gradually, you begin to wonder whether there are any completely uninteresting numbers at all, sitting there unnoticed like wallflowers at the numbers ball.

Can you prove it? Well, yes you can if you approach the question in the manner of many mathematical arguments. You start by assuming that the opposite is true and then use that assumption to deduce something that contradicts it. This means that your first assumption must have been false. It is an ultimate version of the gambit in chess wherein a player offers a piece to an opponent in the knowledge that, if it is taken, it opens the way for a much bigger gain in the future. It is the ultimate version of this manoeuvre because it is the whole game, rather than simply a single piece, that is being offered by this logical gambit.

Let us assume that there are uninteresting positive whole numbers and collect them together. If such a collection exists, then it will have a smallest member. But that smallest member is by definition interesting: it is the smallest uninteresting number. This contradicts our initial assumption that it is an uninteresting number. So our first assumption, that there are uninteresting numbers, was false. All numbers must be 'interesting'.

Just to prove it, here is a story, well known among mathematicians, that is told about the English mathematician Godfrey Hardy when he went to visit his friend, the remarkable Indian mathematician Srinivasa Ramanujan, in a London hospital: 'In the taxi from London, Hardy noticed its number, 1729. He must have thought about it a little because he entered the room where Ramanujan lay in bed and, with scarcely a hello, blurted out his disappointment with it. It was, he declared, "rather a dull number", adding that he hoped that wasn't a bad omen. "No, Hardy," said Ramanujan, "it is a very interesting number. It is the smallest number expressible as the sum of two cubes in two different ways."* Such numbers are now known as 'taxicab' numbers in memory of this incident.

---

* Because $1729 = 1^3 + 12^3 = 9^3 + 10^3$. The cubes have to be positive in these examples. If you allow negative numbers to be cubed, then the smallest such number is $91 = 6^3 + (-5)^3 = 4^3 + 3^3$.

90

Incognito

The number you have reached does not exist.

Recorded message for unobtainable Italian phone numbers*

In the sixteenth and seventeenth centuries it was not uncommon for the leading mathematicians of the day to publish their discoveries in code. This seems very strange to modern scientists, who clamour for recognition and priority in being the first to discover something, but there was some method to the seeming madness of those early mathematicians. They wanted to have their proverbial cake and eat it. Publishing a discovery that makes use of a new mathematical 'trick' establishes you as its discoverer, but it also reveals the trick so that others can use it and beat you to other, possibly greater, discoveries. You have a choice: hold off announcing the first discovery until you have given yourself time to investigate other possibilities more closely – and run the risk that someone else will discover and publish your first result – or you could publish an encoded version of your discovery. Assuming no one breaks the code, your new trick is safe from exploitation by others, and if anyone else comes along and announces they have discovered what you have already found, you can just apply the decrypt to show that you made the discovery long before. Very devious – and I hasten to add that such behaviour does not go on

* In Italian: 'Il numero selezionato da lei è inesistente.'

today in science and mathematics, and probably wouldn't be toler-
ated if it was attempted. However, it does happen in the world of
literature. Books like the political novel *Primary Colours*, about the
Presidential election campaign of Bill Clinton, written by an initially
unidentified journalist using a pseudonym, look a little like an
attempt to have it both ways.

Suppose that you wanted to conceal your identity in a similar
way today, how could you use simple maths to do it? Pick a couple
of very large prime numbers, for example 104729 and 105037
(actually you want to pick much larger ones, with hundreds of
digits, but these are big enough to get the general idea). Multiply
them together to get the product: 11000419973. Incidentally, don't
trust your calculator, it probably can't cope with such a large
number and will round the answer off at some stage – my calculator gave the wrong answer 11000419970.

Now, let's go back to our secret publishing challenge. You want
to publish your discovery and not reveal your identity publicly, but
include a hidden 'signature' so that at some future date you can
show that you wrote it. You could publish the book with that huge
product of two prime numbers, 11000419973, printed in it at the
back. You know the two factors (104729 and 105037) and can easily
multiply them together to show that the answer is your product
'code'. However, if other people start with 11000419973, they will
not find it so easy to find the two factors. If you had chosen to
multiply together two very large prime numbers, each with 400
digits, then the task of finding the two factors could take a
lifetime, even if assisted by a powerful computer. It is not impossible to break our 'code', but using much longer numbers is an
unnecessary level of security – it just needs to take a very long
time.

This operation of multiplying and factoring numbers is an
example of a so-called 'trapdoor' operation (see Chapter 27). It's
quick and easy to go in one direction (like falling through a trapdoor), but longwinded and slow to go in the opposite direction

(like climbing back up through the trapdoor). A more complicated version of multiplying two prime numbers is used as the basis for most of the world's commercial and military codes today. For example, when you buy anything online and enter your credit card details into a secure website, the details are compounded with large prime numbers, transmitted to the company and then decrypted by prime number factorisation.

91

The Ice Skating Paradox

After finishing dinner, Sidney Morgenbesser decides to order dessert. The waitress tells him he has two choices: apple pie and blueberry pie. Sidney orders the apple pie. After a few minutes the waitress returns and says that they also have cherry pie, at which point Morgenbesser says, 'In that case I'll have the blueberry pie!'

Academic legend

When we make choices or cast votes, it seems rational to expect that, if we first chose K as the best among all the alternatives on offer, and then someone comes along and tells us that there is another alternative, Z, which they forgot to include, our new preferred choice will be to stick with K or to choose Z. Any other choice seems irrational because we would be choosing one of the options we rejected first time around in favour of K. How can the addition of the new option change the ranking of the others?

The requirement that this should not be allowed to happen is so engrained in the minds of most economists and mathematicians that it is generally excluded by fiat in the design of voting systems. Yet, we know that human psychology is rarely entirely rational and there are situations where the irrelevant alternative changes the order of our preferences, as it did with Sidney Morgenbesser's pie order (of course, he could have seen one of the pies in question between orders).

A notorious example was a transport system that offered a red bus service as an alternative to the car. Approximately half of all travellers were soon found to use the red bus; half still used a car. A second bus, blue in colour, was introduced. We would expect one quarter of travellers to use the red bus, one quarter to use the blue bus, and one half to continue travelling by car. Why should they care about the colour of the bus? In fact, what happened was that one third used the red bus, one third the blue bus, and one third their car!

There is one infamous situation where the effect of irrelevant alternatives was actually built into a judging procedure, with results so bizarre that they eventually led to the abandonment of that judging process. The situation in question was the judging of ice skating performances at the Winter Olympics in 2002, which saw the young American Sarah Hughes defeat favourites Michelle Kwan and Irina Slutskaya. When you watch skating on the television the scores for an individual performance (6.0, 5.9, etc.) are announced with a great fanfare. However, curiously, those marks do not determine who wins. They are just used to order the skaters. You might have thought that the judges would just add up all the marks from the two programmes (short and long) performed by each individual skater, and the one with the highest score wins the gold medal. Unfortunately, it wasn't like that in 2002 in Salt Lake City. At the end of the short programme the order of the first four skaters was:

Kwan (0.5), Slutskaya (1.0), Cohen (1.5), Hughes (2.0).

They are automatically given the marks 0.5, 1.0, 1.5 and 2.0 because they have taken the first four positions (the lowest score is best). Notice that all those wonderful 6.0s are just forgotten. It doesn't matter by how much the leader beats the second place skater, she only gets a half-point advantage. Then, for the performance of the long programme, the same type of scoring system operates, with the only difference being that the marks are doubled, so the first placed skater is given score 1, second is given 2, third 3 and so on. The score from the two performances are then added together to give each skater's total score. The lowest total wins the gold medal.

After Hughes, Kwan and Cohen had skated their long programmes Hughes was leading, and so had a long-programme score of 1, Kwan was second so had a score of 2, and Cohen lay third with a 3. Adding these together, we see that before Slutskaya skates the overall marks are:

1st: Kwan (2.5), 2nd: Hughes (3.0), 3rd: Cohen (4.5)

Finally, Slutskaya skates and is placed second in the long programme, so now the final scores awarded for the long programme are:

Hughes (1.0), Slutskaya (2.0), Kwan (3.0), Cohen (4.0).

The result is extraordinary: the overall winner is Hughes because the final scores are:

1st: Hughes (3.0), 2nd: Slutskaya (3.0), 3rd: Kwan (3.5), 4th: Cohen (5.5)

Hughes has been placed ahead of Slutskaya because, when the total scores are tied, the superior performance in the long programme is used to break the tie. But the effect of the poorly constructed rules is clear. The performance of Slutskaya results in the positions of Kwan and Hughes being changed. Kwan is ahead of Hughes after both of them have skated, but after Slutskaya skates Kwan finds herself behind Hughes! How can the relative merits of Kwan and Hughes depend on the performance of Slutskaya? The paradox of irrelevant alternatives rules ok.

92

The Rule of Two

History is just one damn thing after another.

Henry Ford

Infinities are tricky things and have perplexed mathematicians and philosophers for thousands of years. Sometimes the sum of a never-ending list of numbers will become infinitely large; sometimes it will get closer and closer to a definite number; sometimes it will defy having any type of definite sum at all. A little while ago I was giving a talk about 'infinity' that included a look at the simple geometric series

$$S = \tfrac{1}{2} + \tfrac{1}{4} + \tfrac{1}{8} + \tfrac{1}{16} + \tfrac{1}{32} + \tfrac{1}{64} + \ldots$$

And so on, forever: every term in the sum exactly half the size of its predecessor. The sum of this series is actually equal to 1, but someone in the audience who wasn't a mathematician wanted to know if there was any way to show him that this is true.

Fortunately, there is a simple demonstration that just uses a picture. Draw a square of size 1 × 1, so its area is 1. Now let's divide the square in half by dividing it from top to bottom into two rectangles. Each of them must have an area equal to ½. Now divide one of these rectangles in two to make two smaller rectangles, each with area equal to one-quarter. Now divide one of these smaller rectangles in half to make two more rectangles, each of area equal to one-eighth. Keep on going like this, making a rectangle of half the area of the previous one, and look at the picture. The original square has just had its whole area subdivided into a never-ending sequence of regions that fill it completely. The total area of the square is equal to the sum of the areas of the pieces that I have left intact at each stage of the cutting process, and the areas of these pieces is just equal to our series S. So the sum of the series S must be equal to 1, the total area of the square.

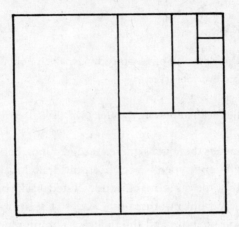

Usually when we encounter a series like S for the first time, we work out its sum in another way. We notice that each successive term is one half of the previous one and then multiply the whole series by 1/2 so we have

$$\tfrac{1}{2} \times S = \tfrac{1}{4} + \tfrac{1}{8} + \tfrac{1}{16} + \tfrac{1}{32} + \tfrac{1}{64} + \ldots$$

But we notice that the series on the right is just the original series, S, minus the first term, which is 1/2. So we have that 1/2 × S = S − 1/2, and S = 1 again.

93

Segregation and Micromotives

The world is full of obvious things which nobody by any chance ever observes.

Sherlock Holmes in *The Hound of the Baskervilles*

In many societies there is a significant segregation between communities of different types – racial, ethnic, religious, cultural and economic. In some cases there is loudly stated dislike of one community by another, but in others there doesn't seem to be any overt attempt to be separate and the different communities get on well as individuals in their spheres of activity. However, the tendencies of individuals may not be a good guide to the behaviour of a group because of the interplay between many individual choices. When some of the statistical methods used by scientists to study the collective behaviour of large numbers of things are applied to populations of people, some very simple but unexpected truths emerge.

In 1978 Thomas Schelling, an American political scientist, decided to investigate how racial segregation came about in American cities. Many people assumed that it was simply a result of racial intolerance. Some thought that it might be overcome by throwing different communities together in a random mix, but were surprised to find that the result was always segregation into different racial sub-communities again, even though the residents seemed to be quite tolerant of other ethnic groups when questioned in surveys. What emerged from a mathematical study of virtual societies using computer simulations was that very slight imbalances result in complete segregation despite an average tolerant outlook. Suppose that a family will move, because of intolerance or to avoid it, if more that one in three of its neighbours are different from them, but will stay put if fewer than one in five are different. In this situation a random mix of two sorts of families ('blue' and 'red') that differ in some way (race, religion or class, say) will gradually become more and more polarised, until eventually it is completely segregated into a totally blue and a totally red community, just like a mixture of oil and water,* with empty 'buffer' regions between them. In a region with above average reds the blues move, leading to above average blues elsewhere, so the reds in that new above average blue neighbourhood move out, and so on. The moves all tend to be towards regions where there is an above average concentration of the mover's type. The boundary regions between different regions are always very sensitive since single movers can tip the balance one way or the other. It is more stable to evolve towards these boundary regions being empty to create a buffer between the segregated communities.

These simple insights were very important. They showed that very strong segregation was virtually inevitable in mixed communities and didn't imply that there was serious intolerance. Segregation doesn't necessarily mean prejudice – although it certainly can, as the examples of the United States, Rhodesia, South Africa and Yugoslavia show. Better to foster close links between the separate communities than to try to prevent them forming. Macrobehaviour is fashioned by micromotives that need not be part of any organised policy.

---

* Actually, this familiar example should not be examined too closely. If the dissolved air is removed from water, say by repeated freezing and thawing, it *will* mix with oil quite smoothly.

94

Not Going with the Flow

Email is a wonderful thing for people whose role in life is to be on top of things. But not for me; my role is to be on the bottom of things.

Don Knuth

We have seen in the previous chapter an example of collective behaviour where no individual wants to find themselves in a significant minority. Not all situations are like this. If you are wanting to get away from it all at an idyllic island holiday destination, you want to be in the minority, not the majority, when it comes to everyone's holiday destination of choice. The pub that 'everyone' chooses to go to because of the music or the food is going to turn out to be a far from ideal experience if you have to queue to get in, can't find a chair and have to wait an hour to be served. You will do better at a less popular venue.

This is like playing a game where you 'win' by being in the minority. Typically, there will be an average number of people who choose to go to each of the venues on offer, but the fluctuations around the average will be very large. In order to reduce them and converge on a strategy that is more useful, it is necessary to use past information about the attendance at the venue. If you just try to guess the psychology of fellow customers, you will end up committing the usual sin of assuming that you are not average. You think that your choice will not also be made by lots of other people acting on the same evidence – that's why you find that everyone else has decided to go for the same stroll by the river on a sunny Sunday afternoon.

If there are two venues to choose between, then as a result of everyone's accumulated experience the optimal strategy gets closer and closer to half of the people going to each venue – so neither is specially popular or unpopular – on the average. At first, the fluctuations around the average are quite large, and you might turn up at one venue to find a smaller than average crowd. As time goes on, you use more and more past experience to evaluate when and if these fluctuations will occur and act accordingly, trying to go to the venue with the smaller number. If everyone acts in this way, the venue will maintain the same average number of attendees over time but the fluctuations will steadily diminish. The last ingredient of this situation is that there will be players who trust their memories and analyses of past experience, and there will be others who don't or who only appeal to experience a fraction of the times when they have to make a choice. This tends to split the population into two groups – those who follow past experience totally and those who ignore it. Since the consequences of making the wrong choice are far more negative (no dinner, wasted evening) than the consequences of making the right choice are positive (quicker dinner, more comfortable evening), greater care is made to avoid wrong choices, and players tend to hedge their bets and go for each available choice with equal probability over the long run. Adopting a more adventurous strategy results in more extreme losses than gains. The result is a rather cautious and far from optimal pattern of group decision making and all the eating venues are less than full.

95

Venn Vill They Ever Learn

> There are two groups of people in the world; those who
> believe that the world can be divided into two groups of
> people, and those who don't.
>
> Anon.

John Venn came from the east of England, near the fishing port
of Hull, and went – as promising mathematicians did – to
Cambridge, where he entered Gonville and Caius College as a
student in 1853. Graduating among the top half-dozen students in
mathematics, he was elected into a college teaching fellowship.
He then left the college for four years and was ordained a priest
in 1859, following in the line of his distinguished father and grandfather, who were prominent figures in the evangelical wing of the
Anglican Church. However, instead of following the ecclesiastical
path that had been cleared for him, he returned to Caius in 1862
to teach logic and probability. Despite this nexus of chance, logic
and theology, Venn was also a practical man and rather good at
building machines. He constructed one for bowling cricket ball
which was good enough to clean bowl one of the members of
the Australian touring cricket team on four occasions when they
visited Cambridge in 1909.

It was his college lectures in logic and probability that made
Venn famous. In 1880 he introduced a handy diagram for representing logical possibilities. It soon replaced alternatives that had

been tried by the great Swiss mathematician Leonard Euler and the Oxford logician and Victorian surrealist writer Lewis Carroll. It was eventually dubbed the 'Venn diagram' in 1918.

Venn's diagrams represented possibilities by regions of space. Here is a simple one that represents all the possibilities where there are two attributes.

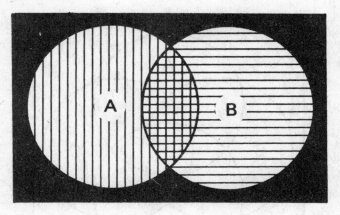

Suppose A is the collection of all brown animals and B is the collection of all cats. Then the hatched overlap region contains all brown cats; the region of A not intersected by B contains all the brown animals other than cats; the region of B not intersected by A represents all the cats that are not brown; and finally, the black region outside both A and B represents everything that is neither a brown animal nor a cat.

These diagrams are widely used to display all the different sets of possibilities that can exist. Yet one must be very careful when using them. They are constrained by the 'logic' of the two-dimensional page they are drawn on. Suppose we represent four different sets by the circles A, B, C and D. They are going to represent the collections of friendships between three people that can exist among four people Alex, Bob, Chris and Dave. The region A represents mutual friendships between Alex, Bob and Chris; region B friendships between Alex, Bob and Dave; region C friendships between Bob, Chris and Dave; and region D friendships between Chris, Dave and Alex. The way the Venn-like diagram has been drawn displays a sub-region where A, B, C and D all intersect. This would mean that the overlap region contains someone who is a member of A, B, C and D. But there is no such person who is common to all those four sets.

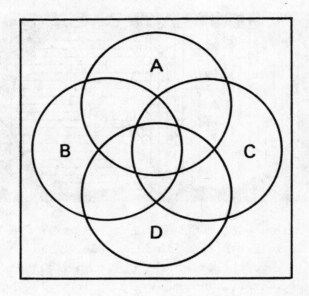

96

Some Benefits of Irrationality

Mysticism might be characterised as the study of those propositions which are equivalent to their own negations. The Western point of view is that the class of all such propositions is empty. The Eastern point of view is that this class is empty if and only if it isn't.

Raymond Smullyan

There is more to photocopying than meets the eye. If you do it here in Europe, then you will soon appreciate a feature so nice that you will have taken it for granted. Put two sheets of A4 paper side by side face-down on the copier and you will be able to reduce them so that they print out, side by side, on a single piece of A4 paper. The fit of the reduced copy to the paper is exact and there are no awkward extra margins on the page in the final copy. Try to do that in the United States with two pieces of standard US letter-sized paper and you will get a very different outcome. So what is going on here, and what has it got to do with mathematics and irrationality?

The International Standard (I.S.O.) paper sizes, of which A4 is one, derive from a simple observation first made by the German physicist, Georg Lichtenberg, in 1786. Each paper size in the so called A-series has half the area of the next biggest sheet because it is half as wide but just as long. So putting two sheets side by side creates a sheet of the next size up: for example, two A4 sheets make one A3 sheet. If the length is L and the width is W, this means that they must be chosen so that $L/W = 2W/L$. This requires $L^2 = 2W^2$, and so the lengths of the sides are in proportion to the square root of 2, an irrational number that is approximately equal to 1.41: $L/W = \sqrt{2}$.

This irrational ratio between the length and the width of every paper size, called the 'aspect ratio' of the paper, is the defining feature of the A paper series. The largest sheet, called A0, is defined to have an area of 1 square metre, so its dimensions are $L(A0) = 2^{1/4}$ m and $W(A0) = 2^{-1/4}$ m, respectively. The aspect ratio means that a sheet of A1 has length $2^{-1/4}$ and width $2^{-3/4}$, so its area is just ½ square metre. Continuing this pattern, you might like to check that the dimensions of a piece of AN paper, where $N = 0,1,2,3,4,5, \ldots$ etc., will be

$$L(AN) = 2^{1/4-N/2} \text{ and } W(AN) = 2^{-1/4-N/2}$$

The area of a single sheet of AN paper will therefore be equal to the width times the length, which is 2^{-N} square metres.

All sorts of aspect ratios other than $\sqrt{2}$ could have been chosen. If you were that way inclined you might have gone for the Golden Ratio, so beloved by artists and architects in ancient times. This choice would correspond to picking paper sizes with $L/W = (L+W)/L$, so $L/W = (1+\sqrt{5})/2$, but it would not be a wise choice in practice.

The beauty of the $\sqrt{2}$ aspect ratio becomes most obvious if we return to the photocopier. It means that you can reduce one side of A3, or two sheets of A4 side by side, down to a single sheet of A4 without leaving a gap on the final printed page. You will notice that the control panel on your copier offers you a 70% (or 71% if it is a more pedantic make) reduction of A3 to A4. The reason is that 0.71 is approximately equal to $1/\sqrt{2}$ and is just right for reducing one A3 or two A4 sheets to one of A4. Two dimensions, L and W, are reduced to $L\sqrt{2}$ and $W\sqrt{2}$, so that the area LW is reduced to $LW\sqrt{2}$, as required if we want to reduce a sheet of any AN size down to the size below. Likewise, for enlargements, the number that appears on the control panel is 140% (or 141% on some photo-copiers) because $\sqrt{2} = 1.41$ approximately. Another consequence of this constant aspect ratio for all reductions and magnifications is that diagrams retain the same relative shapes: squares do not become rectangles and the circles do not become ellipses when their sizes are changed between A series papers.

Things are usually different in America and Canada. The American National Standards Institute (ANSI) paper sizes in use there, in inches because that is how they were defined, are A or Letter (8.5 in × 11.0 in), B or Legal (11 in × 17 in), C or Executive (17 in × 22 in), D Ledger (22 in × 34 in), and then E Ledger (34 in × 44 in). They have two different aspect ratios: alternately 17/11 and 22/17. So, if you want to keep the same aspect ratio when merging paper sizes you need to jump two paper sizes rather than one. As a result, you cannot reduce or magnify two sheets of one size down to one sheet of the size below or above without leaving some empty margin on the copy. When you want to make reduced or enlarged copies on a US photocopier you have to change the paper trays around in order to accommodate papers with two aspect ratios rather than using the one $\sqrt{2}$ factor that we do in the rest of the world. Sometimes a little bit of irrationality helps.

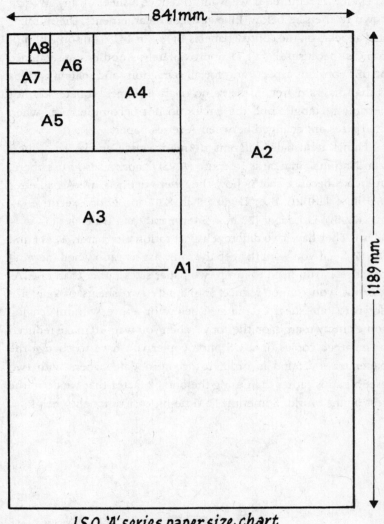

ISO 'A' series paper size chart

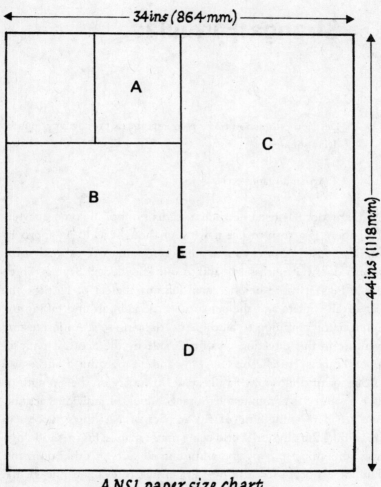

ANSI paper size chart

97

Strange Formulae

Decision plus action times planning equals productivity minus delay squared.

Armando Iannucci

Mathematics has become such a status symbol in some quarters that there is a rush to use it without thought as to its appropriateness. Just because you can use symbols to re-express some words does not necessarily add to our knowledge. Saying 'Three Little Pigs' is more helpful than defining the set of all pigs, the set of all triplets and the set of all little animals and taking the intersection common to all three overlapping sets. An interesting venture in this direction was first made by the Scottish philosopher Francis Hutcheson in 1725, and he became a successful professor of philosophy at Glasgow University on the strength of it. He wanted to compute the moral goodness of individual actions. We see here something of the impact of Newton's success in describing the physical world using mathematics: his methodology was something to copy and admire in all sorts of other domains. Hutcheson proposed a universal formula to evaluate the virtue, or degree of benevolence, of our actions:

$$\text{Virtue} = \frac{\text{Public Good} \pm \text{Private Interest}}{\text{Natural Ability to do Good}}$$

Hutcheson's formula for the moral arithmetic has a number of pleasing features. If two people have the same natural ability to do good, then the greatest one that produces the largest public good is the more virtuous. Similarly, if two people produce the same level of public good then the one of lesser natural ability is the more virtuous.

The other ingredient in Hutcheson's formula, Private Interest, can contribute positively or negatively ($\pm$). If a person's action benefits the public but harms themselves (for example, they do charitable work for no pay instead of taking paid employment), the Virtue is boosted by Public Good + Private Interest. But if their actions help the public and also themselves (for example, campaigning to stop an unsightly property development that blights their own property as well as their neighbours') then the Virtue of that action is diminished by the factor Public Good – Private Interest.

Hutcheson didn't attribute numerical values to the quantities in his formula but was prepared to adopt them if needed. The moral formula doesn't really help you because it reveals nothing new. All the information it contains has been plugged in to create it in the first place. Any attempt to calibrate the units of Virtue, Self-Interest and Natural Ability would be entirely subjective and no measurable prediction could ever be made. None the less, the formula is a handy shorthand for a lot of words.

Something strangely reminiscent of Hutcheson flight of rationalistic fantasy appeared 200 years later in a fascinating project embarked upon by the famous American mathematician George Birkhoff, who was intrigued by the problem of quantifying aesthetic appreciation. He devoted a long period of his career to the search for a way of quantifying what appeals to us in music, art and design. His studies gathered examples from many cultures and his book *Aesthetic Measure* makes fascinating reading still. Remarkably, he boils it all down to a single formula that reminds me of Hutcheson's. He believed that aesthetic quality is determined by a measure that is determined by the ratio of order to complexity:

Aesthetic Measure = Order/Complexity

He sets about devising ways to calculate the Order and Complexity of particular patterns and shapes in an objective way and applies them to all sorts of vase shapes, tiling patterns, friezes and designs. Of course, as in any aesthetic evaluation, it does not make sense to compare vases with paintings: you have to stay within a particular medium and form for this to make any sense. In the case of polygonal shapes Birkhoff's measure of Order adds up scores for the presence or absence of four different symmetries that can be present and subtracts a penalty (of 1 or 2) for certain unsatisfactory ingredients (for example, if the distances between vertices are too small, or the interior angles too close to 0 or 180 degrees, or there is a lack of symmetry). The result is a number that can never exceed 7. The Complexity is defined to be the number of straight lines that contain at least one side of the polygon. So, for a square it is 4, but for a Roman Cross (like the one shown here) it is 8 (4 horizontals plus 4 verticals):

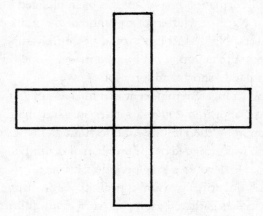

Birkhoff's formula has the merit of using numbers to score the aesthetic elements, but unfortunately, aesthetic complexity is too broad for such a simple formula to encompass and, unlike

Hutcheson's cruder attempt, it fails to create a measure that many would agree on. If one applies his formula to modern fractal patterns that appeal to so many (not just mathematicians) with their repeating patterns on smaller and smaller scales, then their order can score no more than 7 but their complexity becomes larger and larger as the pattern explores smaller and smaller scales and its Aesthetic Measure tends rapidly to zero.

98

Chaos

Other countries have unpredictable futures but Russia is a
country with an unpredictable past.

Yuri Afanasiev

Chaos is extreme sensitivity to ignorance. It arises in situations
where a little bit of ignorance about the current state of affairs
grows rapidly as time passes, not merely increasing in proportion
to the number of time-steps taken but roughly doubling at each
step. The famous example of this sort is the weather. We fail to
predict the weather in England with high accuracy on many occasions, not because we are not good at predicting or because there
is some special unknown secret of meteorology that physicists
have failed to discover, but because of our imperfect knowledge
of the state of the weather now. We have weather stations every
100 kilometres over parts of the country, fewer over the sea, which
take periodic measurements. However, this still leaves scope for
considerable variations to exist between the weather stations. The
Met Office computer has to extrapolate using a likely form of in-
between-the-weather-stations weather. Unfortunately, little differences in that extrapolation often lead to very different future
weather conditions.

This type of sensitivity of the future to the present began to
be studied extensively in the 1970s when inexpensive personal
computers started to be available to scientists. It was dubbed 'chaos'

to reflect the unexpected outcomes that could soon result from seemingly innocuous starting conditions because of the rapid growth in the effects of any small level of uncertainty. The movie business latched on to this with the film *Jurassic Park*, where a small mistake, which led to cross-breeding with dinosaurs, and a broken test tube led to a disaster, as things rapidly went from bad to worse and a small amount of uncertainty ballooned into almost total ignorance. There was even a 'chaologist' on hand to explain it all as we watched the problems snowball out of control.

There is one interesting aspect of chaos that resonates with entirely non-mathematical experiences that we have of books, music and drama. How do we evaluate whether a book or a play or a piece of music is 'good' or better than some other? Why do we think that *The Tempest* is better than *Waiting for Godot*, or Beethoven's 5th Symphony superior to John Cage's *4' 33"*, which consists of 4 minutes and 33 seconds* of silence?

One way is to argue that good books are the ones we want to read again, good plays make us want to see them again, and good music creates a desire to hear another performance of the same work. We seek to do that because they possess a little bit of chaotic unpredictability. A little change in the direction and cast of *The Tempest*, a different orchestra and conductor or a different state of mind in us the reader, will result in a very different overall experience of the play, the music or the book. Run of the mill art lacks that quality. Changes of circumstance produce much the same overall experience. There is no need to have that experience again. Chaos is not just something to be avoided or controlled.

Some people think that the possibility of chaos is the end of

---

* I have always been surprised to discover that my 'revelation' as to why it is 4 minutes and 33 seconds (and no other number) is news to all the musicians who have talked to me about it. In fact, the interval of time of 273 seconds was chosen by Cage for his Absolute Zero of sound by analogy with the minus 273 degrees Celsius which is the Absolute Zero of temperature where all classical molecular motion ceases. Remarkably, Cage once claimed this was his most important work.

science. There must always be some level of ignorance about everything in the world – we don't have perfect instruments. How can we hope to predict or understand anything if there is fast growth in these uncertainties. Fortunately, even though individual atoms and molecules move chaotically in the room where I am sitting, their average movement as a whole is entirely predictable. Many chaotic systems have this nice property and we actually use some of those average quantities to measure what is happening. Temperature, for example, is a measure of the average speed of the molecules in the room. Even though the individual molecules have a history that would be impossible to predict after a few collisions with their neighbours and other denser objects in the room, these collisions keep the average fairly steady and eminently predictable. Chaos is not the end of the world.

99

All Aboard

I am not a cloud bunny, I am not an aerosexual. I don't like
aeroplanes. I never wanted to be a pilot like those other
platoons of goons who populate the air industry.

Michael O'Leary, boss of Ryanair

If you have spent as much time as I have queuing to get on aeroplanes, then you know all the bad plans that can be made. The
budget airlines, like Ryanair, just don't care: it's a free for all with
no seat reservations. Then they realised there was actually an incentive to make it as bad as possible for a while, so that they could
sell you the right to 'priority board' ahead of the rest. There is no
special priority for those with small children or mobility problems,
so these passengers slow the general boarding even more. What
happens when everyone elects to priority board? I don't know, but
I suspect it is the ultimate aim of the idea.

Commercial airlines have a variety of methods to alleviate the
stress and reduce the delay for economy passengers. Everyone has
an assigned seat, and children and those needing extra time get to
board first. Some airlines will board by seat number, so that when
there is a single entrance at the front, those seated at the back get
on first and don't obstruct other passengers needing to pass them.
It all sounds fine, on paper, but in practice somebody blocks the
aisle as they try to get their oversize luggage into the overhead
rack, people sitting in the aisle seats have to keep getting up to let

in those sitting by the windows, and everyone is in somebody's way. There has got to be a better system.

A young German astrophysicist, Jason Steffen, based at Fermi Lab near Chicago, thought the same, and started to explore the efficiency of different passenger loading strategies using a simple computer simulation that could accommodate changes in boarding strategy and add in many random variations that perturb the best-laid plans. His virtual aircraft has 120 seats, accommodating 6 passengers per row divided by a central aisle and has no business or first class section. All the virtual passengers have hand baggage to stow.

It was easy to find the worst loading policy to adopt for a plane with a single entrance at the front: load by seat number starting from the front. All passengers have to fight their way past the passengers already on board and busy packing their hand baggage in order to get to their seats. This is fairly obvious, but it led airlines to conclude that the best strategy is simply the opposite of the worst one: load by seat number from the back. Remarkably, Steffen's investigation found that this was actually the second slowest method of boarding! Only boarding from the front was worse. Even boarding completely at random, irrespective of seat number, did much better. But the best method was more struc-tured. Passengers should board so that those sitting at the windows go ahead of those sitting in the middle or on the aisles and they should board in a way that distributes the number of people trying to stow luggage at the same time along the whole length of the plane rather than congregate all in one area.

If passengers in all even-numbered rows with window seats board first, they have a clear row of aisle space in front and behind and don't get in each others' way while stowing bags. Everyone can load bags at the same time. If anyone needs to pass, then there are spare aisles to step into. By beginning from the back, the need to pass other passengers is avoided again. Those sitting in the middle and aisle seats follow on. Then the passengers in the odd-numbered rows follow.

Not everyone can easily follow this strategy to the letter. Small children need to stay with their parents, but they can still board first. However, the time gained by making it the basic strategy could be considerable. The computer model showed that over hundreds of trials with different small variations ('problem passengers'), this method was on the average about seven times quicker at loading passengers than the standard load-from-the-back-method. Steffen has patented it!

100

The Global Village

Imagine all the people
Sharing all the world.

John Lennon

Sometimes you can't see the wood for the trees. Large numbers
swamp our understanding. It's hard to imagine a million, let alone
a billion. Cutting things down to size helps us to make things more
concrete and immediate. In 1990 a famous illustration of the state
of the World was suggested,* which asked us to imagine that the
World's population was scaled down to create a single village of
100 people, with all their attributes scaled down accordingly. What
would that village be like?

It would contain 57 Asians and 21 Europeans, 14 members from
the Western hemisphere, both north and south, and just 8 Africans.

---

* This picture of a global village was first proposed in 1990 by Donatella Meadows
and was based on a population of 1,000. After hearing Meadows talking on the
radio, the environmental activist David Copeland tracked down Meadows and
revised her statistics to reflect a village population of 100. He added it to a poster
prepared for distribution to 50,000 people at the 1992 Earth Summit in Rio.
Meadows's original example of the State of the Village report was published in
1990 as *Who Lives in the Global Village?* A myth grew up that the originator of this
example was a Stanford University professor, Philip Harter, but in fact he only
forwarded an email of the Meadows and Copeland facts over the internet. See
http://www.odtmaps.com/behind_the_maps/population_map/state-of-village-
stats.asp

There would be 70 villagers who were non-white and 30 who were white. Only 6 people would possess 59% of the entire village's wealth and all 6 would be from the United States. Of the village's 100 members, 80 would live in substandard housing, 70 would be unable to read, 50 would suffer from malnutrition, only one would own a computer, and only one would have had a college education. A strange village, isn't it?

Notes

'Immersing myself in a book or a lengthy article used to be easy. My mind would get caught up in the narrative or the turns of the argument, and I'd spend hours strolling through long stretches of prose. That's rarely the case any more. Now my concentration often starts to drift after two or three pages. I get fidgety, lose the thread, begin to look for something else to do. I feel as if I'm always dragging my wayward brain back to the text. The deep reading that used to come naturally has become a struggle.'

Nicholas Carr

1   The chain is assumed to be of uniform mass per unit length, is perfectly flexible and has zero thickness. The mathematical function $\cosh(...)$ is defined in terms of the exponential function by $\cosh(x) = (e^x + e^{-x})/2$.

2   Assume the maximum acceleration you can exert is $+A$, and hence the maximum deceleration is $-A$, and you have to push the car from position $x = 0$ to $x = 2D$. You start at zero speed and finish at zero speed. If you apply $+A$ for the distance from $x = 0$ to $x = D$, then it will take a time $\sqrt{(2D/A)}$ to get to $x = D$ and you will arrive there with speed $\sqrt{(2DA)}$. If you immediately apply deceleration $-A$ to the car then it will return to zero speed at $x = 2D$ after a period of deceleration lasting a time $\sqrt{(2D/A)}$. As expected by the symmetry of the situation, this is exactly equal to the time taken to reach halfway. The total time to get the car in the garage is therefore $2\sqrt{(2D/A)}$.

3   The random walk is a diffusion process governed by the diffusion equation. For the spread of y in time, t, and one dimension of distance, x,

this is $\partial y/\partial t = K\partial^2 y/\partial x^2$, where K is a constant measure of how easy it is to diffuse through the medium. Hence, by inspecting the dimensions we have that y/t should be proportional to y/x^2. Hence, t is proportional to x^2 and we will need S^2 steps to go the straight-line distance x = S.

4   This length is known as the Planck length to physicists, after Max Planck, one of the pioneers of quantum theory, who first determined it. It is the only quantity with the dimensions of a length that can be formed from the three great constants of nature, c the speed of light, h, the quantum constant of Planck, and G, the constant of gravitation. It is equal to $(Gh/c^5)^{1/2}$ and is a unique reflection of the relativistic, quantum and gravitational character of the universe. It is a unit of length that is not picked out by human convenience and this is why it seems so small in terms of our everyday units.

5   A simple example is the monkey puzzle where you have, say, 25 square pieces that each have 4 sides. On each side is half of a monkey (top or bottom). There are four colours of monkey and you have to put the 25 pieces together into a larger 5 × 5 square so that each border completes a monkey's body by joining a top to a bottom of the same colour. How many alternative ways of putting the cards down would you have to search through in order to find the correct 'answer' to the puzzle in which all the half monkeys are joined to other pieces of the same colour that match the other part of their body? There are 25 ways to put down the first piece, followed by 24 ways for the next one, 23 ways for the next, and so on. Hence, there are 25 × 24 × 23 × 22 ... × 3 × 2 × 1 = 25! ways to match up the 25 cards in this way. But there are 4 possible orientations for each card and this creates another 4^{25} possibilities. The total number of configurations to be tried in order to find the correct pattern is therefore 25! × 4^{25}. This number is stupendously large. If we tried to write it out in full it would not fit on the page. If a computer were to search through this huge number of moves at a rate of one million per second it would still take more than 5,333 trillion trillion years to examine them all in order to find the right answer. For comparison, our Universe has been expanding for only 13.7 billion years.

6   Note that if the best candidate is in the (r+1)st position and we are skipping the first r candidates, then we will choose the best candidate for sure, but this situation will occur only with a chance 1/N. If the best candidate is in the (r+2)st position we will pick them with a chance 1/N × (r/r+1). Carrying on for the higher positions, we see that the overall

probability of success is just the sum of all these quantities, which is $P(N,r)$
$= 1/N \times [1 + r/(r+1) + r/(r+2) + r/(r+3) + r/(r+4) + r/(r+5) + \ldots + r/(N-1)] \approx 1/N \times [1 + r \ln[(N-1)/r]$. This last quantity, which the series converges towards as N gets large, has its maximum value when the logarithm $\ln[(N-1)/r] = 1$, so $e = (N-1)/r \approx N/r$ when N is large. Hence the maximum value of $P(N,r)$ occurring there is $P \approx r/N \times \ln(N/r) \approx 1/e \approx 0.37$.

7   More precisely, if we skip the first N/e applicants then our probability of finding the best one will approach $1/e$ as N becomes large,
where $e = 2.7182 \ldots = 1/0.37$, is the mathematical constant that
defines the base of the natural logarithms.

8   The chance that one person will not have your birthday is $364/365$ and
so, if there are G guests and their birthdays are independent, the probability that none of them will share your birthday is $P = (364/365)^G$.
Therefore the chance that there will be one of the guests who shares
your birthday will be $1 - P$. As G gets larger and larger P approaches zero
and the chance of sharing your birth date with another guest approaches
1, and we can check that $1 - P$ exceeds 0.5 when G exceeds $\log(0.5)/\log(364/365)$ which is approximately 253.

9   Again, it is easiest to start by working out the probability that they don't
share birthdays. If there are N people then this probability equals $P = 365/365 \times 364/365 \times 363/365 \times \ldots \times [365 - (N-1)]/365$. The first term
is from allowing the first person to choose their birthday freely (any one
of the 365 days is allowed). The second term is then the fraction allowed
for the second person if they are not to have the same birthday as the
first – so they can only have 364 of the 365 possibilities. Then the third
person can be on only 363 of the 365 days to avoid having the same
birthday as either of the first two, and so on, down to the Nth person
who can only choose 365 minus (N-1) of the 365 days to avoid having
the same birthday as any of the other N-1 before him. So the chance
that two of them *do* share a birthday is just $1 - P = 1 - N!/\{365^N(365 - N)!\}$ which is bigger than 0.5 when N exceeds 22.

10 If you alter the confidence probability from 95% to some other
number, say P%, then the three intervals in the diagram have lengths
$\frac{1}{2} \times (1-P/100)$, $P/100$ and $\frac{1}{2} \times (1-P/100)$, respectively (check the
previous answer is obtained when you put $P = 95$). Following the
same reasoning as before, we see that at A the future is $[P/100 + \frac{1}{2} \times (1-P/100)]/\frac{1}{2} \times (1-P/100) = [100 + P]/[100 - P]$ times longer than
the past. So, with P% probability something will last at least $[100 -$

P]/[100 + P] times its present age, but at most [100 + P]/[100 − P] times that age. As P gets close to 100% the predictions get weaker and weaker because they have to be so much more certain. If we want 99% probability then it will last more than 1/199 times the present age but less than 199 times that age. On the other hand, if we let the certainty of being right drop to, say, 50% then the likely interval is 1/3 to 3 times the present age: a very narrow interval but the chance of it being correct is too low to worry about.

11 A Maclaurin series is a series approximation to a function f of one variable x by a series of polynomial terms. Tamm would write $f(x) = f(0) + xf'(0) + \ldots x^n f^{(n)}(0)/n! + R_n$ where R_n is the error, or remainder, after approximating $f(x)$ by n of the preceding terms, where n = 1,2,3, . . . as required. The remainder term that Tamm was asked to deduce can be calculated to be $R_n = \int_o^x x^n\, f^{(n+1)}(t)/n!\, dt$. If the bandit leader was particularly awkward he might have required the further simplification of this expression that is possible by using the mean value theorem to obtain $R_n = x^{n+1}\, f^{(n+1)}(y)/(n+1)!$, where y is some value between 0 and x. Colin Maclaurin was a Scottish contemporary of Isaac Newton.

12 This approximation is best for very small values of the interest rate r. If we want to make it more accurate, then we can approximate $\log_e(1+r)$ by $r - r^2/2$ and obtain the estimate that $n = 0.7/r(1 - \tfrac{1}{2}r)$. For the case of 5% interest rates, with r = 0.05, this gives the number of years for the investment to double as 14.36 instead of 14 years.

13 This is a beautiful application of what is known as the 'method of dimensions' in physics. Taylor wants to know the radius, R, of a spherical explosion at a time t after it occurs (call that detonation time t=0). It is assumed to depend on the energy, E, released by the bomb and the initial density, ρ, of the surrounding air. If there exists a formula $R = kE^a\rho^b t^c$, where k, a, b and c are numbers to be determined, then, because the dimensions of energy are ML^2T^{-2}, of density are ML^{-3}, where M is mass, L is length and T is time, we must have a = 1/5, b = −1/5 and c = 2/5. The formula is therefore $R = kE^{1/5}\rho^{-1/5}t^{2/5}$. Assuming that the constant k is fairly close to 1, we see that the unknown energy is given approximately by $E = \rho R^5/t^2$. By comparing several pictures you can determine k as well.

14 It's an elephant!

15 Any 3-digit number ABC can be written out as 100A + 10B + C. The first step is to take away its reverse, that is 100C + 10B +A. The result

is $99 | A-C |$, where the straight brackets mean take the magnitude with a positive sign. Now the magnitude of A-C must lie between 2 and 9 and so when you multiply it by 99 you get one of the multiples of 99 that has three digits. There are only 8 of these that are possible answers: the numbers 198, 297, 396, 495, 594, 693, 792 and 891. Notice that the middle digit is always a 9 and the other two digits always add up to 9 in each of these numbers. Therefore, when you add any one of them to its reverse you always end up with the answer 1089.

16 If the Leader spoke the truth with probability p then the probability that his statement was indeed true is $Q = p^2/[p^2 + (1-p)^2]$. In our case, where $p = \frac{1}{4}$, this gives a probability of $Q = \frac{1}{10}$. You can check that Q is less than p whenever $p < \frac{1}{2}$ and Q exceeds p whenever $p > \frac{1}{2}$. Notice that when $p = \frac{1}{2}$ that $Q = \frac{1}{2}$ also.

17 For a more detailed discussion see J. Haigh, *Taking Chances*, Oxford UP, Oxford (1999), chap. 2. The essential quantity required to work out any result is that the odds for matching r of the drawn numbers is 1 in $^{49}C_6/[^6C_r {}^{42}C_{6-r}]$ where $^nC_r = n!/(n-r)!r!$ is the number of different combinations of r different numbers that can be chosen from n numbers.

18 If you create the Fibonacci sequence of numbers (in which each number from the third onwards is the sum of the two before it), 1,1,2,3,5,8,13,21,34,55,89 and so on forever, with the nth number in the sequence labelled as F_n, then you can generalise the example here by rearranging an $F_n \times F_n$ square into an $F_{n-1} \times F_{n+1}$ rectangle. The example we have drawn is the n = 6 case as $F_6 = 8$. The area difference between the square and the rectangle is given by Cassini's identity, $(F_n \times F_n) - (F_{n-1} \times F_{n+1}) = (-1)^{n+1}$. So when n is an even number the right-hand side equals -1 and the rectangle is bigger than the square; but when n is an odd number we 'lose' an area of 1 when we go from the square to the rectangle. Remarkably the difference is always +1 or -1 no matter what the value of n. This puzzle was invented by Lewis Carroll and became popular in Victorian England. See S. D. Collingwood, *The Lewis Carroll Picture Book*, Fisher Unwin, London (1899) pp. 316–7.

19 For more information see Donald Saari, 'Suppose You Want to Vote Strategically', *Maths Horizons*, November 2000, p. 9.

20 Tolkowsky showed that for total internal reflection of incoming light to occur at the first face it strikes, the inclined facets must make an angle of more than 48° 52' with the horizontal. After the first internal

reflection the light will strike the second inclined face and will be completely reflected if the face makes an angle of less than 43° 43' with the horizontal. In order for the light to leave along a line that is close to vertical (rather than close to the horizontal plane of the diamond face) and for the best possible dispersion in the colours of the outgoing light, the best value was found to be 40° 45'. Modern cutting can deviate from these values by a small amount in order to match properties of individual stones and variations in style.

[21] That is for any two quantities x and y it is always true that $\frac{1}{2}$ (x+y $\geq$ $(xy)^{1/2}$, which is a consequence of the fact that the square $(x^{1/2}-y^{1/2})^2$ ≥ 0. This formula also displays an attractive feature that the geometric mean possesses that the arithmetic mean does not. If the quantities x and y were measured in different units ($ per ounce and £ per kilogram, for instance) then you could not compare the value of $\frac{1}{2}$ (x+y) at different times meaningfully. To do that, you need to make sure that x and y are both expressed in the same units. However, the geometric mean quantity $(xy)^{1/2}$ can be compared in value at different times even if different units are being used to measure x and y.